Praise for *Super Fly*

"Flies! Those irritating insects that settle on your food when you eat outside in summer, cluster round the eyes of horses, carry diseases on their little tickling feet. How can someone write a whole book on flies! The best thing I can say is: Read *Super Fly*! It is utterly fascinating, written with clear prose, a delightful sense of humor, by a gifted naturalist and storyteller. And Jonathan Balcombe writes not only with authority about the incredible diversity of fly species, but with a real love for these fascinating winged beings that play such an important role in the tapestry of life."

—Jane Goodall, PhD, DBE, founder, Jane Goodall Institute, and UN Messenger of Peace

"Read this engaging and well-researched book and learn why we can't live without flies and other insects. Then understand and respect their 'ecological services' and wonder about what and how they may feel with many senses and abilities far more developed than our own."

—Michael W. Fox, veterinarian, ethologist, and author of *Animals and Nature First*

"Ogden Nash wrote, 'God in His wisdom made the fly, and then forgot to tell us why.' Now Jonathan Balcombe's witty book enlightens us, advising of the fly's, and other insects', surprising role in preserving our ecosystem and far more. In my view, the first thoroughly readable, enjoyable, and scholarly work on the subject."

—Ingrid Newkirk, cofounder and president, People for the Ethical Treatment of Animals (PETA)

"True to form, Jonathan Balcombe's deep interest in flies nicely follows his work on the behavior and cognitive and emotional lives of fishes, nonhumans whom numerous people think of as merely edible streams of protein. In *Super Fly*, Balcombe clearly shows that flies are complex and wonderful beings—not disposable or swattable pests who are dumb and unfeeling but rather individuals whose lives matter to them and whose existence should and must matter to us. I can only hope that when people get done reading this highly unique, important, and fact-filled book they will show flies and other marginalized animals the respect they truly deserve. We can learn a lot about ourselves by peering into the remarkable lives of these remarkable insects."

—Marc Bekoff, coauthor of *The Animals' Agenda* and *A Dog's World: Imagining the Lives of Dogs in a World Without Humans*

"Written with infectious passion and a large dose of empathy, *Super Fly* is bound to astonish and delight you. Combining science with storytelling, and clarity with grace and humor, Balcombe shows a willingness to go where others have been hesitant to venture. I cannot recommend this book highly enough."

—Jeffrey Moussaieff Masson, PhD, coauthor of *When Elephants Weep: The Emotional Lives of Animals*

"Just when you thought humans were the dominant animal on the planet, Jonathan Balcombe swoops in with his characteristically entertaining prose to remind us that for each one of us, there are actually 17 million flies. Yet how much do we know about these ubiquitous and important creatures? After reading their riveting story here, you'll not only cure yourself of Diptera ignorance, but you'll have the most interesting stories to tell at any party you attend."

—Paul Shapiro, author of *Clean Meat: How Growing Meat without Animals Will Revolutionize Dinner and the World*

"Our planet is home to over 160,000 species of flies, from microscopic midges to giant robber flies that can take down a hummingbird—wingless flies, flies that swim underwater, bloodsucking flies, flies that live in rhinoceros stomachs. Combining meticulous research with superb storytelling, *Super Fly* covers every aspect of the behavior, biology, and impact on humanity of creatures that are annoying, deadly, and fascinating. This book cements Jonathan Balcombe's status as one of today's best science writers, and it will make you think twice the next time you pick up a flyswatter."

—Hal Herzog, author of *Some We Love, Some We Hate, Some We Eat: Why It's So Hard to Think Straight about Animals*

"About any topic at all, Jonathan Balcombe is a fluid and engaging writer, and I have devoured his previous books. This one does not disappoint, offering an entertaining tour of a highly accomplished group of minibeasts. Read it, learn, and maybe find yourself empathizing in ways you would have thought impossible."

—Bruce Friedrich, cofounder and executive director, The Good Food Institute

"Jonathan Balcombe has long been a respected voice for the 'other animals,' providing us with insightful and empathetic views of the inner lives of mostly misunderstood corners of the animal kingdom. In

Super Fly he again combines his skills as a researcher with his mastery of the narrative to expose the inner workings of the most ubiquitous order of animals, the true flies."

—Stephen A. Marshall, professor of entomology, University of Guelph, Ontario, and author of *Flies: The Natural History and Diversity of Diptera*

"Imagine a talented writer who learns almost everything there is to know about flies. Now envision the writer organizing those learnings into a readable and digestible summary, into something that renders flies not only relevant but fascinating, and that, in the end, inspires thoughts that go far beyond the book's stated topic. And with this you have Jonathan Balcombe's *Super Fly*, a celebration of life viewed through a group of insects as delightfully complex as anyone could hope to encounter."

—Bill Streever, biologist and bestselling author of *Cold* and *In Oceans Deep*

"We go through life unaware of the incredibly diverse and abundant world of insects that surround us. Too often, insects are vilified. Flies especially get a bad rap. Think flies and we tend to focus on malaria, yellow fever, and cholera. And yet, we could not survive without flies. In *Super Fly*, animal behaviorist Jonathan Balcombe zooms in on the fascinating world of flies like no one else has. Jonathan writes in such an engaging and often humorous manner, I never thought I would say this about a book on flies: I loved this book!"

—Aysha Akhtar, MD, MPH, author of *Our Symphony with Animals: On Health, Empathy, and Our Shared Destinies*

Praise for *What a Fish Knows*

"We Buddhists consider all animals, including fish, as sentient beings who have feelings of joy and pain just as we humans do. We also believe that they have all been kind to us as our mothers many times in the past, and are deserving of our compassion. Therefore, we try to help them in whatever way we can and at least avoid doing them harm. In *What a Fish Knows*, Jonathan Balcombe vividly shows that fish have feelings and deserve consideration and protection like other sentient beings. I hope reading it will help people become more aware of the benefits of vegetarianism and the need to treat animals with respect."

—The Dalai Lama

"An extended exploration of the world from a piscine perspective . . . Balcombe makes a persuasive case that what fish know is quite a lot."
—Elizabeth Kolbert, *The New York Review of Books*

"[An] exhaustively researched and elegantly written argument for the moral claims of ichthyofauna." —Nathan Heller, *The New Yorker*

"*What a Fish Knows* will leave you humbled, thrilled, and floored. Jonathan Balcombe delivers a revelation on every page, presenting jaw-dropping studies and stories that should reshape our understanding of, and compassion for, some of the most diverse and successful animals who have ever lived. After reading this, you will never be able to deny that fishes love their lives as we love ours, and that they, too, are vividly emotional, intelligent, and conscious. Bravo!"
—Sy Montgomery, author of *The Soul of an Octopus*, a National Book Award finalist

"Balcombe builds a persuasive argument. Writing in a straightforward, somewhat breezy style, he makes his case partly through a compendium of fascinating anecdotes and scientific findings that illustrate the complexity and creativity of fish behavior. . . . Dozens of startling revelations emerge." —Alan de Queiroz, *The Wall Street Journal*

"I thought I knew a lot about fishes. Then I read *What a Fish Knows*. And now I know a lot about fishes! Stunning in the way it reveals so many astonishing things about the fishes who populate planet Earth in their trillions, this book is sure to 'deepen' your appreciation for our fin-bearing co-voyagers, the bright strangers whose world we share."
—Carl Safina, author of *Beyond Words*

PENGUIN BOOKS

SUPER FLY

Jonathan Balcombe was born in England and has lived in New Zealand, the United States, and Canada. A biologist with a PhD in ethology, the study of animal behavior, he is the author of four popular science books on the inner lives of animals—the *New York Times* bestselling *What a Fish Knows, The Exultant Ark, Second Nature,* and *Pleasurable Kingdom*—as well as more than sixty scientific papers and book chapters on animal behavior and animal protection. Formerly the department chair for animal studies with the Humane Society University and the director of animal sentience with the Humane Society Institute for Science and Policy, he has also served as an associate editor for the journal *Animal Sentience* and has taught a course in animal sentience for the Viridis Graduate Institute. Balcombe currently lives in southern Ontario, where in his spare time he enjoys biking, baking, birding, Bach, and trying to understand the squirrels in his neighborhood.

Super Fly

The Unexpected Lives of the World's Most Successful Insects

JONATHAN BALCOMBE

PENGUIN BOOKS

PENGUIN BOOKS
An imprint of Penguin Random House LLC
penguinrandomhouse.com

Grateful acknowledgment is made for permission to reprint the following:
Excerpt from "Blackfly Song" by Wade Hemsworth, copyright © 1963 by Southern Music Publishing Co. Inc. Copyright © renewed. International rights secured. Used by permission. All rights reserved.
"The Fly" by Ogden Nash, copyright © 1942 by Ogden Nash, renewed. Reprinted by permission of Curtis Brown Ltd.

LIBRARY OF CONGRESS CATALOGING-IN-PUBLICATION DATA
Names: Balcombe, Jonathan P., author.
Title: Super fly : the unexpected lives of the world's most successful insects / Jonathan Balcombe.
Description: New York : Penguin Books, [2021] |
Includes bibliographical references and index.
Identifiers: LCCN 2020019489 (print) | LCCN 2020019490 (ebook) |
ISBN 9780143134275 (paperback) | ISBN 9780525506041 (ebook)
Subjects: LCSH: Flies. | Flies—Behavior. | Flies—Adaptation.
Classification: LCC QL531 .B35 2020 (print) | LCC QL531 (ebook) |
DDC 595.77—dc23
LC record available at https://lccn.loc.gov/2020019489
LC ebook record available at https://lccn.loc.gov/2020019490

Printed in the United States of America
1st Printing

Set in Absara Pro
Designed by Sabrina Bowers

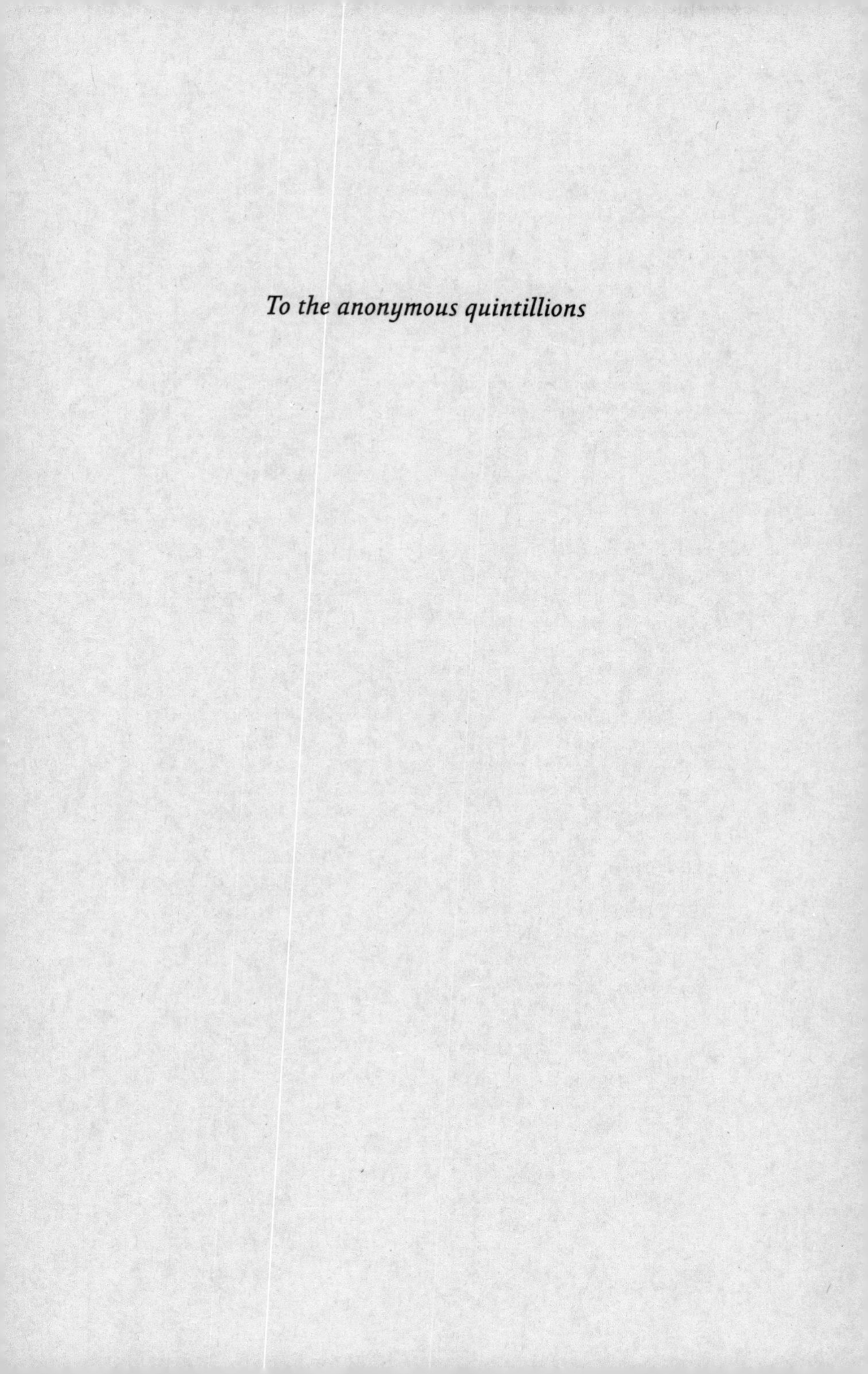

To the anonymous quintillions

Contents

Part I

What Flies Are

Chapter 1

God's Favorite

> Human knowledge will be erased from the world's archives before we possess the last word that a gnat has to say to us.
>
> —JEAN-HENRI FABRE

On about the sixth day, I realized that the four tiny red welts on my chest were not mosquito bites. It was our third week of a month's sojourn in Kruger National Park, South Africa, where I was one of a team of 14 biologists studying the movements and roosting habits of bats. A small group of us were taking a lunch break during a foray on foot to track the locations of several radio-tagged African yellow house bats.

I had noticed that the welts were becoming larger and itchier with each passing day but had shrugged it off, thinking I must be more sensitive to the bites of whatever African mosquito had had its way with me. As I absentmindedly scratched the bumps through my shirt between bites of a sandwich, I felt a strange sensation—a faint tickling. I peeled off my shirt and scrutinized one of the welts.

It was moving.

Years earlier, I had read of large botfly maggots tunneling their way under the skin of the arms and legs of a teenager who had miraculously survived a midair plane explosion en route to Lima in the 1970s. Her earthward plummet was cushioned by vegetation, and she awoke, still strapped into her seat, in the Amazon jungle. Armed with courage, determination, and a knowledge of edible plants she had learned from her botanist parents, she survived a twelve-day solo hike through the bush to civilization.

My infestation was less dramatic. These were not botflies. Back at camp, Leo Braack, our South African park ranger who happened to be an authority on parasitic flies, soon identified my uninvited guests as African skin maggots, *Cordylobia anthropophaga*. *Anthropophaga* translates to "man-eater." Drawn to the rank odor of sweaty clothing I had hung up to dry, the mother fly had laid her eggs on the unclean garments, and when I had redonned them thinking I could get away with a second wearing, the maggots, stirred by my body heat, had emerged to tunnel through my skin. Burrowed headfirst into my flesh, the hungry grubs breathed through a minuscule hole at the surface. My four tiny wounds were painless, but itchy.

I should say that while the label *man-eater* is technically true, this was not the sort of consumption that has given certain sharks and tigers their ill-deserved reputations. I wasn't about to lose a limb or spill blood. Nevertheless, it is disconcerting to discover another creature gnawing away at your flesh, however small. Suddenly, my own lunch was displaced by a new priority concerning someone else's: *I wanted them out!*

An hour later, while I posed for photos at our campsite along the Luvuvhu River, Braack instructed me on how to remove the maggots:

"Just rub a little Vaseline over the opening, and you'll be able to squeeze them out in about 30 minutes."

"That's comforting," I thought. "Easy for you to say."

I retired to a shady spot with a tube of Vaseline and a good

The author poses while park ranger Leo Braack documents a range extension for the African skin maggot in Kruger Park, South Africa.

(© BROCK FENTON)

book. An hour later I had expressed three pearly-white, rice-grain-size maggots. The fourth one held out until the next day.

Not only was I the only human on the trip to host African skin maggots, but I was at that time the only person in history to host them at our location, according to a delighted Braack. Common though they are, African skin maggots had not been recorded that far south on the African continent. I was soon being lovingly referred to as "the ecosystem" by my comrades, and I became the butt of hygiene humor for the remainder of the trip. Apparently, none of them caught the irony that I, the only vegetarian in the whole group, should be the one whose flesh was deemed most suitable for consumption by a fly.

Unpopular and Important

Let's face it, flies do not win popularity contests with us. Among our most feared animals, flies are vastly outranked by the likes of spiders, snakes, lions, and crocodiles. But if one were to survey humankind for animals we most dislike, flies would make many top-ten lists. "Of all the major groups of insects, the flies are the least understood and most detested," writes entomologist Mark Deyrup in his 1999 book, *Florida's Fabulous Insects*. "There are no apologists for flies, there are no lobbyists or hobbyists for flies, there are no fly-watchers, no fly-gardens, no picture guides to flies." (Deyrup's last claim has been rendered obsolete, as we'll soon see.) For sheer repugnance to humans, an adult fly is surely trumped by its fellow insect the cockroach, but a juicy fly maggot propelling itself across the putrid flesh of a rotting carcass with successive undulations of its viscera visible through translucent skin makes for stiff competition on the yuck scale.

Then there is their dastardly lust for blood. While most of us go through life without playing host to a flesh-eating maggot, it is a rare human who has not suffered the unsettling whine of a mosquito or felt the familiar itch of her bite. Chances are that anyone reading this will also have been harassed by blackflies, sand flies, deerflies, stable flies, and/or horseflies. Having spent thousands of hours exploring the North American outdoors, I have been targeted by all of these aerial phlebotomists. The business end of a large horsefly has a set of mouthparts that work like alternating saws to penetrate the skin, and the pain is nothing to scoff at. I was terrified by my first encounter with one while swimming at an Ontario summer camp as a young boy. The big black creature swooped down onto the heads of swimmers when they surfaced. The pain of its bite was immediate and severe. I desperately wanted to turn into a fish. I once saw a large horsefly gorging on the flank of a cow in the Texas Hill Country, blood dripping copiously from the wound.

If painful bites were the only cost of cohabiting this planet with dipterans, we'd have it good. Flies wreak far greater havoc as vectors of deadly tropical diseases unwittingly delivered to humans and other animals through their bites. Half of all clinical cases of disease in the world are transmitted by insects, and flies are the most common carriers. A human dies of malaria every 12 seconds, and mosquitoes of various species are its primary couriers. Not ones to retreat from mayhem, mosquitoes also deliver the microbes that cause yellow fever, dengue, Zika, filariasis, and encephalitis.

Mosquitoes are not the only culprits. Tropical sand flies spread leishmaniasis in humans, and tropical blackflies can carry the roundworm that causes river blindness. One in six humans alive today is infected by an insect-borne illness, and more often than not, the footprint left at the crime scene is that of a fly.

I am not writing this book to demonize flies. I have no personal grudge against them. Only a tiny proportion, about 1 percent, of the 160,000 known species of flies are harmful to humans. By contrast, the beneficial and pretty flower flies (Syrphidae) alone, which are vital pollinators, number over 6,000 described species. Our common antipathy toward insects in general, and flies in particular, obscures a range of critical beneficial services they perform, including pollination, waste removal, natural pest control, and being a critical food source for scores of other animals. Few of us are aware of these and other fly benefits. For instance, you probably didn't know (I didn't) that midge larvae around the world are an important antipollution brigade; in their multitudes—billions per acre in some locales—they filter algae and microscopic debris from the water, which they draw in a little stream through an open-ended tube they build around themselves, facedown in the mud. Even the devilish bites of certain flies have hidden benefits—that is, if you don't adhere too closely to anthropocentrism. Biting flies have kept humans out of ecologically sensitive areas, preventing habitat and biodiversity loss. Case in point: the lush Okavango Delta of Botswana—a seasonal floodplain spanning some 16,800 square kilometers (6,500 square miles)—is a paradise for wildlife and a stronghold of the tsetse fly, whose bite can sicken both humans and their cattle.

Flies also play major roles in science. Modern genetics owes much to the fruit fly, *Drosophila melanogaster*, which has been the subject of over a hundred thousand published studies.* And crime solving owes a debt to Diptera. Such is the speed and efficiency with which certain flies colonize our dead bodies that entomologists, armed with an intimate knowledge of the life history of these fly species, can determine time of death to within a few hours. This technique has aided hundreds of murder convictions, and exonerations.

Megadiverse

Useful or not, flies are hugely successful. I did not choose the subtitle of this book lightly, nor am I hedging with the claim that they rank as "God's favorite."

What do I mean by the *success* of flies? That adjective seems hardly apt for an imprisoned housefly bouncing inanely against a windowpane. What I am actually referring to is a more biologi-

*As of February 12, 2020, a search under *Drosophila* on the National Library of Medicine's PubMed database yields 107,760 hits.

cal sort of success: diversity and sheer numbers. On these terms, flies' success takes on celestial proportions.

First off, flies belong to by far the most successful collection of animals on Earth: insects. "It is easy to forget that human beings form a tiny two-legged minority in an overwhelmingly six-legged world," writes Canadian entomologist Stephen Marshall in the introduction to his 2006 book, *Insects*. Insects make up a whopping 80 percent of the approximately 1.5 million animal species so far named, and there are estimated to be between 5 and 10 million species yet to be discovered. At any one time, there are some ten quintillion (10,000,000,000,000,000,000) insects crawling, hopping, burrowing, boring, or flying. That's 200 million for every living human, according to *Animal Life Encyclopedia* author Bernhard Grzimek. In his 2017 book *Bugged*, journalist David MacNeal presents an even more skewed scoresheet: 1.4 billion insects for every human. Ants alone are thought to outweigh humans twelve times over, and Lisa Margonelli reports in her book *Underbug* that termites outweigh us by a similar ratio. A typical backyard may contain several thousand species of insects and several million individuals.

Nobody knows how many living flies there are on planet Earth at any one time, but researchers at the Animalist channel think there are about 17 quadrillion (17,000,000,000,000,000). British fly expert Erica McAlister estimates there are about 17 million flies for every human. With numbers like these, you may rightly wonder why we are not constantly mobbed by clouds of pesky gnats, mosquitoes, and crane flies. The reason is that most flies are in preadult stages (eggs, larvae, or pupae) and thus lack the conspicuous characteristic they are named for. Nonetheless, such is the abundance and ubiquity of flies that, as you read this, you're likely within a few feet, if not a few inches, of some sort of fly. Wherever you are in the world, if the weather is warm and you spend time outdoors, you will almost certainly be in physical contact with at least one fly on any given day.

You may be excused for doubting the above numbers. It is hardly as though the air and the ground are swarming with insects. But there are vast expanses of land, particularly in far northern latitudes, where insects at their reproductive peak, and flies especially, really do swarm in prodigious number. The Russian translator of one of my books sent me links to videos of tens of thousands of horseflies and blackflies swarming on and around an all-terrain vehicle in a Siberian wetland. The videographers are well protected in netting and gloves, but I shudder for any reindeer who treads there. Then there are the midges, which might turn out to be the most dominant collection of species on Earth. Phil Townsend, a remote-sensing specialist at the University of Wisconsin in Madison, reported in 2008 the laying down of 135 kilograms of dead midges per hectare (120 pounds per acre) per day around Iceland's Lake Mývatn (English translation: Midge Lake). Some phantom midges amass in such huge numbers in East Africa that locals catch them in swinging buckets, then pack them into balls and cook them into edible masses called kungu cakes.

For the record, I'm not suggesting that any kind of fly is the most abundant species on Earth. As we go to smaller organisms, some of their numbers rise astronomically. There are more living organisms in a single teaspoon of healthy soil than there are people on Earth. One of the most abundant animals on the planet is a well-studied nematode (roundworm) called *Coenorhabditis elegans*. A British biologist estimated that 600 quintillion of them are born every day. According to a 1998 estimate, there are about 5×10^{30} bacteria on this planet.

Another measure of evolutionary success is number of species. Depending on which expert you ask, flies may rank first, second, or third (after beetles and maybe wasps/ants/bees) as the most species-rich order of animals on Earth. In the 1930s, British geneticist J. B. S. Haldane famously said that God had "an inordinate fondness for beetles," owing to beetles' fantastic diversity, which at that time far outranked that of flies. Today there are about one million known species of insects, of which 350,000 are beetles. But most flies are generally more elusive and obscure than most beetles, and as scientists have redoubled their efforts and honed their skills at collecting and identifying new species, flies have been catching up.

There were about 80,000 known species when Harold Oldroyd's classic book *The Natural History of Flies* was published in 1964. That number has since doubled to 160,000, and there are signs that we are still only scratching the surface. A DNA barcoding study from 2016 estimated the diversity of gall midges in Canada to exceed 16,000 species—10 times the predicted number. Extrapolating this finding leads to a startling prediction: "If Canada possesses about 1 percent of the global fauna, as it does for known taxa, the results of this study suggest the presence of 10 million insect species with about 1.8 million of these taxa in the gall midge family Cecidomyiidae. If so, the global species count for this fly family could exceed the combined total for all 142 beetle families." Haldane must be rolling in his grave. According to one fly specialist I spoke to, there may be some exaggeration in this extrapolation, but clearly they are "a huge, huge group," nearly entirely undescribed and mostly plant-feeders. At present there are only 6,203 named species of gall midge worldwide.

Steve Marshall is unambiguous in his appraisal of flies' place at the top of the diversity heap. I met Marshall on the suburban campus of the University of Guelph, about an hour's drive west of Toronto, where for 35 years he has served on the environmental biology faculty and as director of the university's world-renowned insect collection. During that time he has built an impressive résumé that includes well over 200 scholarly publications and several magnificent volumes on insect life illustrated with thousands of his own arresting macrophotographs. Alongside Art Borkent (whom we'll meet later), Marshall is Canada's fly guy.

"Something to know about Diptera is that it is probably the most diverse order on the planet," Marshall told me from the other side of the large desk in his office. "In my opinion, the only real challenger in the race to be recognized as the most diverse order is the Hymenoptera [wasps, ants, and bees]."

"Is that a general consensus now?" I asked.

"Coleopterists [beetle scientists] won't agree with it. But I'm confident that there are more species of flies than beetles, even though there are currently almost twice as many named beetle species as there are named fly species."

Marshall's confidence is partly attributable to the rapid rate at which new fly species are now being discovered. To illustrate his point, Marshall turned to a graduate student in a corner of his lab who was working on a new batch of flies collected from the neotropics.

"Tiffany, what's the novelty rate in the genus you are currently studying?" he asked. (The novelty rate is the proportion of species found in a sample that are new to science.)

"Ninety to ninety-five percent."

Marshall glanced back at me. "That is from a collection of about six thousand specimens of one genus, from which thirty-seven new species have emerged so far."

Gustavo, another grad student, stood at an adjoining bench poring over specimens of micropezids belonging to the genus *Cardiacephala*.

"Gustavo, what's your novelty rate?"

"It's about fifty percent."

"And that's for big flies pretty easily spotted in the field," continued Marshall. "So even for the most conspicuous flies, one half of the species collected were previously unknown to us."

"Have you ever had a sample with a hundred percent novelty rate?" I asked.

"Oh, yes, especially in the tropics, where a lot of areas remain poorly studied and some genera of smaller flies are made up entirely of new species. When I started here in 1982, even in Guelph the novelty rate for some of the lesser-known families was over half."

Marshall couldn't say off the top of his head how many new fly species he and his team have described and named, but it is certainly over 1,400. It's a laborious and lengthy process, following strict guidelines and requiring very detailed formal descriptions to ensure that the newly named species is distinct from other closely related species.

I wondered aloud: "Do two biologists ever simultaneously describe and name a new species, and if so, how does one deal with that awkward scenario?"

I half expected Marshall to dismiss so unlikely a coincidence, but this is a man of surprises:

"It happened once to me, in 2012. It was a species of *Speolepta*, a North American fungus gnat genus with only one other described species. They live in caves, often on lake shorelines, where they hang threads at the ends of which they pupate. Some of these habits are very similar to those of the famed, predatory cave-dwelling glowworms of New Zealand, which belong to a related group of fungus gnats."

As a child I had experienced the unforgettable sight of tens of thousands of luminescent larvae glowing on the ceiling of the caves at Waitomo, on New Zealand's North Island. It was like gazing up at a clear night sky.

Marshall continued: "The eerie coincidence was not just that we independently described this new species at the same time, but that we gave it the same name! We each named the same species after Richard Vockeroth, one of the all-time greats, and an authority on the genus, who had just recently died."

"How did you find out?" I asked.

"It was six years from the initial discovery before we [the paper was coauthored by Mycetophilidae expert Jan Ševčík] had the paper ready to submit. After we submitted the paper, we found out that Jostein Kjærandsen at Tromsø University Museum, in Norway, had also prepared a paper describing a new species as *Speolepta vockerothi* and, remarkably, it was the same species we'd been working on. We simply invited him to be a coauthor on one shared paper describing the species, and he agreed."

Speolepta vockerothi, a fungus gnat with, as yet, no common name (might I boldly propose Vockeroth's fungus gnat?), made its in-print debut in February 2012 in *The Canadian Entomologist*.

New fly species are being described at the brisk rate of about 1 percent per year, or about 1,600 new species. Because describing and naming new species (taxonomy) is a meticulous and time-consuming pursuit practiced by specialists, the rate of new species entering the books is not limited by fly diversity but by human effort.

As a measure of how esoteric the study of flies can be, and the industriousness of its aficionados, consider that there is a three-volume set, totaling over 1,000 pages, titled *Horseflies of the Ethiopian Region*, which describes 565 species, including 228 new to science at the time of publication in 1957. In the same academic library (Cornell University's Mann Library) where I discovered this gem, I noticed entire volumes on scuttle flies, bee flies, stiletto flies, midges, robber flies, houseflies, snail-killing flies, and of course fruit flies. I happened to open an aged volume titled *Papers on Diptera by CP Alexander, 1910 to 1914*, and there in the frontispiece I found a brief note written and signed by the author: "Presented to the Comstock Memorial Library, Dec. 30, 1914." Alexander (1889–1981) is a legend among entomologists, perhaps the most prolific dipterist who ever lived. During his 60-plus-year career, he described over 11,000 new fly species, an unbelievable pace of about one every two days.

Biotechnology is also accelerating species counts. New DNA barcoding techniques* are revealing vastly greater species diversity than was previously known. A Canadian study from 2016 nearly doubled the number of insect species in Canada, from 54,000 to 94,000. The study also found an unexpected hyperdiversity in one family of flies; more than one in six species of the total were gall midges or gall gnats, very small, wispy flies often less than a millimeter long (20 in a row would not reach an inch).

Fecundity is another measure of an organism's success, or at least its potential for success. Flies make lots of babies, and as we'll see in chapter 8, they have some kinky ways of getting it on. I know of no better illustration of the reproductive potential of flies, indeed insects, than what I read in the introduction to my undergraduate entomology class textbook. It was a hypothetical scenario, thank goodness. The story begins with a pair of fruit flies who mate on January 1. A typical fruit fly brood numbers about 100 eggs, which hatch out as hungry larvae who, all going

*DNA barcoding uses short DNA segments to identify new species or to confirm membership in a known species.

well, stuff themselves on succulent, overripe fruit, pupate, then emerge as a new generation of adults. On average, half of a given brood will be female, and each of those girl flies will produce about 100 babies of her own. Such is the speed with which fruit flies romp through their life cycles that they can complete 25 generations in a year.

Now, casting aside all previous 24 generations, take just the 25th generation of flies emerging from their pupae on December 31 of our hypothetical year. Imagine, then, that these flies are packed into a ball of 1,000 flies per cubic inch.

How big do you think the ball is?

I have posed this question to dozens of people, and invariably they underestimate the size of the ball. Would it be the size of a house? How about the size of a football stadium? Occasionally, someone will propose a ball the size of planet Earth. Bravo to them for thinking outside the box, but still they have come up short, woefully short. The number 50^{24} is no trifling figure. That bulging ball of buzzing beings would have a diameter of 96,372,988 miles, stretching from here to the sun with a few million miles to spare.

Houseflies are scarcely less prolific than fruit flies. In 1911, American entomologist Clinton F. Hodge calculated that a single pair of houseflies, mated in April, could by August, if all their offspring lived, produce over 191,010,000,000,000,000,000 (that's 191 quintillion) adults. If each occupied a 1/8-inch cube, they would cover the Earth to a depth of three stories, in just five months.

These calculations carry another lesson from nature: the critical role of checks and balances. In the real world, only a small proportion of fruit fly eggs survive to become fruit fly maggots, and a tiny fraction of those make it to the pupal stage, of which very few will ever emerge to take wing as breeding adults, who in turn must navigate many perils if they are to successfully contribute to the next round. At each step of the way, nature trims and prunes. In nature's balanced network of food webs, flies couple in fathomable numbers, and their losses along the way fuel the lives of other creatures in the network. When you behold an adult fly, you're looking at a lottery winner.

In addition to being fecund, flies are ubiquitous. While writing this book, many flies dropped in to mark my progress. I was visited by a fruit fly in a library and an unidentified species at a Starbucks coffee shop, drawn predictably to the rim of my mug. On many occasions, and in all seasons, tiny phorid flies darted in frenetic bursts across my computer screen; one got trapped in the deadly viscosity of my glass of water and died despite my best efforts to save it. Away from my desk, I played host to an unknown number of mosquitoes, deerflies, no-see-ums, stable flies, and other winged assailants who met with less sympathy than my phorid. Flies have been loitering in our midst for as long as there have been humans with midsts to loiter in. The first "fly on the wall" surely eavesdropped in a cave.

No continent is too inhospitable for flies. Even Antarctica is home to a few intrepid midges, and a handful of species have colonized the oceans—a habitat otherwise unreached by insects. Some northern midges can dehydrate themselves down to withstand –15°C (5°F) without ice crystals destroying their cell membranes. Other midge larvae live over 1,000 meters (3,200 feet) beneath the surface of Lake Baikal, the world's deepest body of freshwater. Flies' habitats can be hostile, and profoundly obscure. The *Encyclopædia Britannica* observes that there may be no life-supporting medium in which fly larvae have not been found. As their name hints, the larvae of the petroleum fly (*Helaeomyia petrolei*) develop in natural pools of crude oil, where they breathe through a snorkel and feed on the remains of other insects trapped in the goo. Another larva matures in the excretory glands of land crabs. I would never have thought of passing my adolescence in wombat dung, millipede droppings, or the feces of New Zealand's short-tailed bats, but flies did.

Cultured Flies

As a candidate for most bizarre fly habitat, consider cheese. To be precise, a Sardinian sheep milk cheese named casu marzu, which translates into English as "rotten/putrid cheese." You might think a descriptor like that would relegate said cheese to a well-sealed waste receptacle. In fact, the fly's presence—or to be more exact, the maggot's—is indispensable to the particular whiff and taste of this regional delicacy. Larvae of the cheese fly (*Piophila casei*) are deliberately included in the recipe. Over weeks, through a process of digestion and excretion, which may more accurately be described as decomposition and fermentation, the curd matures into a very soft, pungent cheese.

The ⅓-inch-long cheese maggots are notably athletic. Known also as cheese skippers, they can launch themselves over six inches into the air. The maggot does this by using its mouth hooks as grapples to grasp its tail end, then letting go with a snap.* Some

*Research published in 2019 reveals that unrelated gall-midge maggots jump even farther, using a Velcro-like latching system that allows them to leap 36 times their body length to escape danger. "They store elastic energy by forming their body into a loop and pressurizing part of their body to form a temporary 'leg.' They prevent movement during elastic loading by placing two regions covered with microstructures against each other, which likely serve as a newly described adhesive latch." They can do this repeatedly and it is dozens of times more efficient than crawling. See abstract of G. M. Farley et al., "Adhesive Latching and Legless Leaping in Small, Worm-like Insect Larvae,"

diners clear the cheese of its maggots before consumption, and some don't. "As with all maggots, they taste like what they're feeding on," reports one gourmand. Consuming cheese fly maggots is not risk-free. There are confirmed cases of them surviving ingestion and—call them tough—managing to make a living in the host's intestine, a condition called pseudomyiasis, which can result in holes in the intestine accompanied by vomiting, diarrhea, and internal bleeding. Cheese skippers are found worldwide, and they are not especially fussy about their food. In addition to cheeses, they can be found on meats, fatty foods, and decaying bodies.

With habitats like these, it should surprise no one that flies show no reverence. Their brazenness suggests a confidence in their ability to evade harm. The bush fly, an Australian native closely related to our familiar housefly, is so noted for its insolent trespassing on human heads and faces that efforts to repel it have become known as the Australian salute. The proliferation of humans (and cattle) in Australia has been a boon for the bush fly, 100 of which may be bred from a single human stool; in some locales, they occur at densities as high as 9,000 per acre.

To be truly irreverent you must also pay no heed to the elite. If you were attentively watching the run-up to the 2016 US presidential election, you might have noticed a fly land on Hillary Clinton's eyebrow during a presidential debate. It was a cameo appearance, lasting barely a second, but it was enough to spawn slow-motion YouTube clips and the Twitter hashtag #flyforpresident.* President Obama referred to pesky flies during more than

---

Journal of Experimental Biology 222, no. 15 (August 2019), https://jeb.biologists.org/content/222/15/jeb201129 (accessed May 2020).

*As this book was going to press, another fly perched for two minutes on the most conspicuous surface in the entire venue during a televised vice presidential debate: Republican candidate Mike Pence's short-cropped silver hair.

one interview, even joking about them when they barged in on the scene. Athletes, too, get no respect. At the 2018 World Cup soccer tournament, players in the England–Tunisia match were accompanied by clouds of gnats. A 2007 Major League Baseball playoff game was dubbed the Midges Game after the tiny flies descended on the stadium in the eighth inning, altering the outcome and, by some accounts, the series. In August 2018, a fly sabotaged a world-record mini dominoes tumble attempt in Germany, when it landed on one of the fingernail-size stones, causing it to fall and trigger a disastrous cascade. From princes to paupers, nobody is immune to a fly's attentions; they are among life's great equalizers. "Flies and priests can enter any house," goes one Russian proverb.

Flies' presence in dozens of proverbs across many nationalities speaks both to their ubiquity and their cultural presence. Most English speakers will be familiar with the proverbial "fly on the wall" who secretly witnesses all that goes on in its presence. "A fly in the ointment" has lost some of its former popularity, owing perhaps to the fade-out of the word *ointment*. Not so the proverb "A closed mouth catches no flies," an advisory that it is sometimes better to remain silent. Among other things, flies have been proverbialized for fallibility (Every fly has its shadow), vanity (The fly on the back of a water buffalo thinks that it's taller than the buffalo), elusiveness (You can't kill a fly with a spear), overkill (Do not use a hatchet to remove a fly from your friend's forehead), and the power of positivity (It's easier to catch flies with honey than with vinegar).

Flies are not strangers to the visual arts, either. In pre-17th-century Western painting, the presence of a fly on a portrait meant that the subject had died. During the Renaissance, the placement of a trompe l'oeil fly on a canvas became a popular way for artists to showcase their technical skills, especially among Dutch still-life painters.

An example of the symbolic use of flies in art can be seen in *The Discovery of America by Christopher Columbus*, a large painting (almost 14 × 9 feet) by 20th-century Surrealist Salvador Dalí. It includes a depiction of flies in the role of freeing Spain when they emerged from the crypt of Saint Narcisa (whose symbol is the fly) and drove off French invaders. Dalí enhanced the flies' heroic symbolism by morphing them, their wings extended, into crosses. Flies are a symbol of Catalan identity, and Dalí also depicts hundreds of them in a later painting titled *The Hallucinogenic Toreador*. Los Angeles–based artist John Knuth has used flies to produce canvases of color and pattern. Knuth rears hundreds of thousands of houseflies from maggots acquired from commercial suppliers. The adult flies lap up mixtures of water, sugar and watercolor paint provided by Knuth. The flies "paint" by regurgitating tiny blots of liquid, a natural behavior that accompanies feeding. Over the course of months, the colored blots accumulate on a canvas placed in the flies' enclosure, eventually giving rise to atmospheric pointillist creations.

Inevitably, flies have also found expression in song lyrics. While working as a wilderness surveyor in the late 1940s, Canadian songwriter Wade Hemsworth immortalized the blackfly with his eponymous song, which I encountered, along with the fly, at summer camp in my childhood:

> And the black flies, the little black flies
> Always the black fly, no matter where you go
> I'll die with the black fly a-picking my bones
> In North On-tar-i-o-i-o, in North On-tar-i-o

You can listen online to Hemsworth sing the song in a delightful animated short film produced by the Canadian Film Board in 1991. In a sultry 1999 song titled "Last Night of the World," you can hear another Canadian musician, folk/rock icon Bruce Cockburn, "blow a fruit fly off the rim of my glass" while sipping rum in a Guatemalan refugee camp.

Not surprisingly, flies are the source of a goodly amount of humor. According to Groucho Marx, "time flies like an arrow, and fruit flies like a banana." And if you doubt that a maggot can be used to raise one's status, consider what Winston Churchill remarked in 1906 to lifelong friend and confidante Violet Bonham Carter: "We are all worms, but I do believe that I am a glow-worm."

Which brings us to names, and the creative lengths to which scientists have sometimes gone in naming flies. Quasimodo flies are named for an arched thorax, which gives them a hunchbacked appearance. There is a fly genus called *Cinderella*. (A Google search failed to reveal why *Cinderella*, but a kindly fly expert named Norm Woodley did tell me that the name was coined in 1949 from a single specimen collected in Ada, Oklahoma; the anomalous insect couldn't easily be placed in a preexisting fly family, so I wonder if maybe the name refers to Cinderella's poor alignment with her ill-tempered sisters.) There isn't much mystery as to why the names *Calliphora vomitoria* and *C. morticia* were bestowed on flies that colonize decomposing bodies, or why *Elephantomyia* is a genus of crane fly with very long mouthparts. Someone was feeling playful when they named two bee flies for their sounds, *Apolysis humbug* and *A. zzyzxensis*, but the name March fly seems inappropriate for an insect of which few get airborne before April. Perhaps they were anticipating global warming.

Australian entomologist Bryan Lessard (aka Bry the Fly Guy) discovered a new species of fly in a 30-year-old collection box. Because it featured a bright yellow abdomen and was collected in 1981, the year Beyoncé Knowles was born, he named it *Scaptia beyonceae*, the Beyoncé fly.

Flies haven't cornered the market on celebrity monikers. At

least five other insects are named for popular culture icons, including the beetles *Agra katewinsletae*, *Hydroscapha redfordi*, and *Agra liv* for Liv Tyler, and a moth with a striking yellow crown and piercing stare: *Neopalpa donaldtrumpi*.

All naming aside, I have two overarching aims with this book. First, I wish to inspire wonder at the diversity, complexity, and success of a group of animals widely (and in some cases justifiably) disliked, poorly understood, and rarely reflected upon. Second, I hope to raise awareness that our existence on this planet is thanks to a diversity of interacting species, and that, in spite of our antipathy toward them, flies are a vital part of that functioning whole. *Super Fly* explores flies as remarkable opportunists able to make a profitable living in the most unlikely places. I will place flies in human history and culture, relating strange encounters by scientists in the field and homeowners in the kitchen. We will meet flies as gift givers, flesh eaters, lovers, pollinators, blood seekers, predators, parasites and parasitoids, pests, recyclers, deceivers, and cooperators.

I will share the physical prowess of flies: how their wings can beat 1,000 times per second, how their feet stick to windows, how a predaceous robber fly intercepts fast-flying prey on the wing, and how a fly's mouth can act like a syringe (think mosquito), a saw (horsefly), or a sponge (housefly). I will detail the varied bodies and life histories of flies—the frail demureness of the crane fly (see the image in the photo insert); the obscurity of the wingless, parasitic bat fly, which spends its life scurrying around on its furry host; and the audaciousness of the tiny satellite phorid fly, which hovers just beyond reach of the gnashing jaws of a panicked ant, awaiting the opportunity to dash in and inject an egg with its harpoonlike ovipositor (see the image in the photo insert). We will meet dipterists in the field, at their desks, in the lab, and at a professional entomologists' conference.

We'll also meet some spectacular-looking flies: flies with outlandish eye-stalks longer than the rest of their bodies; accomplished mimics that will have you swearing that the fly you're looking at is a bumblebee; minuscule males whose genitalia, relatively speaking, would make a porn star envious; and *holoptic* male flies (see the image in the photo insert), whose enormous cherry-red eyes surround their entire heads like an inflated balloon—all the better to spot a passing female. We will encounter a glittering extravaganza of the improbable, the audacious, and the miraculous ways that flies get on in a world that *seems* to be run only by humans.

I invite you to put away your prejudices against flies, to cast off any layer of angst that may burden your opinions of them, to regard flies nonjudgmentally. If you do, I believe you will, at the very least, marvel at the fantastic diversity of ways that flies have found to make their livings. You might even experience a measure of enchantment and respect toward flies. If you attain that, I will have done my job well. And there's hope in that, for quite simply, we cannot live without them.

Chapter 2

How Flies Work

There are as many organs in a fly as in a leviathan.

—ERNST JÜNGER, *THE GLASS BEES*

Flies' success owes much to their physical makeup. To begin with, flies possess the general characteristics that have made insects the dominant life-form on our planet. As we'll see, they've built some nifty additions onto that framework.

Before we get on to flies, let's briefly visit the basics of how insect bodies function. They work surprisingly like ours.

Evolution is a superb engineer, and insects are marvels of miniaturization. When I regard a tiny midge bobbing up and down in a mating swarm, or a mite (technically not an insect but a close relative)—smaller than the period at the end of this sentence—scurrying across the page of a book, I'm in awe that so much complexity and coordination can be condensed into such a wee package. Insects share eight of the ten body systems we have: nervous, respiratory, digestive, circulatory, excretory, muscular, endocrine, and reproductive. Our two remaining systems—our skeleton and our skin—are replaced in the insect by an exoskeleton made up of rigid plates (*sclerites*) connected by flexible membranes, which provides effective structural support and protection for a small, mobile organism. Like the sections of a symphony orchestra, these systems work in concert.

An open circulatory system moves hemolymph (their equivalent of our blood) throughout the body. Except for when it courses through a dorsal vessel running approximately where our spine would be, the hemolymph flows freely, bathing internal organs, supplying them with oxygen taken in by the ventilatory system, and aiding immunity. Ventilation (or breathing) is handled by a complex, branching set of tubes, the tracheas, which open to the outside of the body at holes called spiracles, whose neat linear arrangement on some large insects resembles the portholes on a boat. Oxygen is absorbed and carbon dioxide removed by direct diffusion at the tracheoles, crossing the outer membrane and entering or exiting the cells within. An active, bellows-like mechanical pumping action of the insect's body makes the exchange more efficient—similar to what our diaphragm does.

Fueling these systems is the food processed by the digestive system, which is arranged much like ours. An insect's foregut substitutes approximately for our stomach, managing food intake and storage. At ingestion, salivary glands lubricate the food and start the process of its digestion. The salivary glands of insects are more versatile than ours; some, for instance, make silk, while others produce compounds that mimic plant growth hormones and stimulate the production of a protective gall—a swelling growth typically occurring on stems or leaves. The midgut, like our small intestine, is where most digestion and absorption of nutrients occurs. Thence to the hindgut, where frass (insect poop) is stored in a muscular rectum and excreted through the anus.

In case you're wondering whether insects fart, they do. Gassy products of lower digestion have to go somewhere, and as for us, the exit door is the anus. Disappointingly, I'm not aware that bug farts are ever audible, but it would not surprise me at all if flies used anal gases to, say, communicate either acoustically or chemically. Herrings communicate by releasing bubbles from their anuses, and bombardier beetles defend themselves against predators with a blast of acid from their rear ends. So if you ever hear a flatulent fly, please let me know.

Every office needs an IT department. An insect's nervous system is a network of neurons anchored to a ventral nerve cord. Along this cord lie nerve centers called ganglia (singular: ganglion). Two principal ganglia lie in the head: (1) the brain, which processes sensory information and is where behavior originates, and (2) the subesophageal ganglion, a dense mass of nerve cells that serve the insect's sensory organs, mouthparts, salivary glands, and neck muscles.

Flyness

Now let's get clear about what flies are. True flies belong to the order Diptera, whose members are distinguished by having just two wings (Greek *di* = two, *ptera* = wings). Flies' ancestral rear wings have been modified into a pair of club-shaped structures called *halteres*, which function chiefly as flight stabilizers. All other flying insects have four functioning wings, with the exception of beetles, whose front wings have been modified into hardened, protective shields called *elytra* (singular: *elytron*).

There are two main groups of flies. The suborder Nematocera includes generally small, delicate flies, such as mosquitoes, crane flies, and midges. Although named for their long antennae (just as *rhinoceros* translates to "nose-horn," so *nematocera* translates to "thread-horns"), you are more likely to identify a nematoceran fly by its slender, frail appearance. The Brachycera include more compact, robust flies with short antennae. The familiar housefly

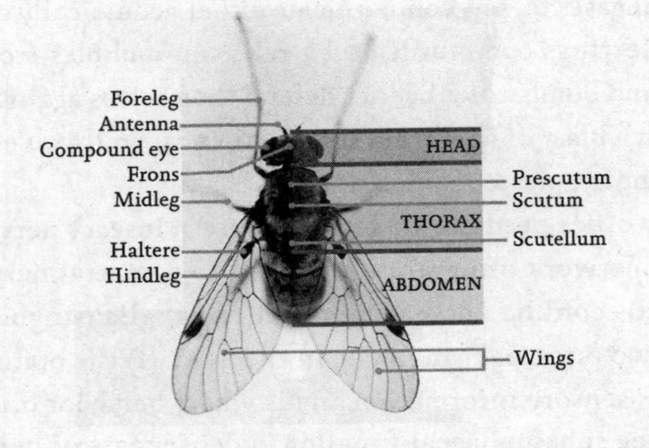

Some parts of a fly.

(© BOB'S BUGS, HTTP://WWW.BOBS-BUGS.INFO/BUG-BASICS-ANATOMY/)

and the blowflies whose maggots found nourishment in my chest tissue belong to the Brachycera.

True to their diversity, flies come in a huge range of shapes and sizes, each exquisitely adapted to its particular lifestyle. Robber flies must be swift and sturdy for aerial predation; the largest grow to nearly three inches (seven centimeters). Several adults of the world's smallest fly (whom we'll meet in chapter 4) would fit on a pinhead, and by my estimation it would take tens of thousands of them to equal the robber fly's mass.

Our antipathy toward flies aligns poorly with their physical beauty. I'll be the first to concede that maggots don't have much going for them aesthetically, but it is the negative associations we have with flies—filth, decay, itching bites, pestilence—that infiltrate our perceptions of these insects. Strip away our cultural angst, and some flies rank among nature's most beautiful works of art: the exquisite symmetry and gleaming metallic color of a blowfly, its celestial blue, green, or gold suit of armor arranged across its thorax and tapering abdomen, gossamer wings glinting, every bristle and wing vein placed and traced as if by the loving hand of a costume designer. At face value, even many of the flies that bite us emerge as objets d'art. The legs of some mosquitoes are adorned with elegant black, brushy leggings (see the image in the photo insert), and the arrangement of facets on the large eyes of some horseflies and deerflies creates psychedelic patterns of light and color. Despite notable competition from butterflies and beetles (members of separate insect orders: Lepidoptera and Coleoptera, respectively), it was a hoverfly, *Metasyrphus americanus*—specifically, a pollen-dusted wasp mimic resting on a flower—that was chosen to grace the cover of volume 3 (*Insects*) of the second edition of the venerable and comprehensive *Grzimek's Animal Life Encyclopedia*.

As part of my quest to know flies, I met entomologist Mark Deyrup at the Archbold Biological Station (ABS) in south-central Florida. Archbold is a 5,193-acre preserve consisting mostly of Florida's distinctive dry scrub habitat. Established in 1941 by the American zoologist and philanthropist Richard Archbold, today the ABS supports over sixty staff and many volunteers. The flora and fauna of this natural gem—which include some of the rarest species in North America—just might be the most thoroughly studied and documented on Earth.

Deyrup, 70, has been at the ABS for 35 years. Energetic enough to pass for 60, he is, like Stephen Marshall, a man who applies his considerable talent and industry to achievements that go largely unheralded by society at large. I first encountered his name when I pulled a copy of his book *Florida's Fabulous Insects* off the shelf while browsing at a local library in nearby Boynton Beach. It is a lively extravaganza of insect life liberally graced by photographs to accompany Deyrup's absorbing prose.

In the spacious lab where he works, Deyrup slid across his

desk a hefty pair of books titled *Manual of Nearctic Diptera*, volumes I and II. Turning to a page at random, I was confronted by exquisite drawings of isolated fly bits. Every bristle had a name (mercifully, the thinner hairs did not). Deyrup pointed to a pair of thoracic bristles on the fly's midsection. "These ones are subapical scutellar setae. Whether they point in parallel like this [I was reminded of a walrus's tusks], or are cruciate like this [a pair of crossed cutlasses] is critical for identification."

There is a reason why insect anatomy and taxonomy texts have extensive arrays of detailed line drawings of obscure structures, bristle patterns, and genitalia: such is the prodigious diversity of flies that some sister species are almost identical. Identifying a fly down to its species, or thereabouts, is done with a detailed road map called a "key." Starting at the coarsest level (example: Is it an insect or not?), a key takes its user through a series of binomial steps, each more specific than the last. All going well, the identification process culminates with a specific feature distinctive to that family, genus, or species. For example, if you have navigated the key correctly and your fly has a spur on the mid-tibial leg segment, then it is a snipe fly from the family Rhagionidae. Steve Marshall's book on flies has a whole section on collecting and preserving Diptera, with ten separate keys for identifying them to the family level.

The use of bristle arrangement for identification is important enough to have a name: *chaetotaxy* (KEY-toe-tax-y). The wings of flies also have a defined pattern of veins, each with a name and characteristic location, often of taxonomic value. If one is trying to identify the maggots, which tend toward few or no bristles, one might focus instead on the arrangement and character of spiracles.

Don't let this detail convince you that we have fly anatomy all figured out. How we perceive a fly and how flies perceive each other may be vastly different. "Nine tenths of the things we see in insects we have no idea what they are there for," Deyrup told me, "because flies are working at a different level. It is just amazing how much is going on with surface chemistry and surface structure in insects, and how little we know about what any of it means."

Frequent Fliers

Flies didn't get their name for nothing. They are master aerialists that can hover, fly backward, and land upside down. It's entirely possible that most of the airborne animals on Earth at any moment are flies. Even beetles, the one group to (currently) outnumber flies in diversity, are more land-prone. If you've spent time handling beetles, you'll have noticed their common reluctance to take flight, in contrast to the hyper-aerial flies.

An insect's small size confers two big advantages for flight, which might explain why insects were flying 150 million years before any other creature was. First, the laws of physics dictate that smaller wings can beat faster. Second, a lighter body is more maneuverable. We can flap our arms about three times in a second, while the smallest bird can reach a hundred beats a second. A housefly attains 345 beats per second; a mosquito up to 700 beats; and a tiny biting midge, a startling 1,046 beats. Paradoxically, faster wingbeats are not just possible for those small wings; they are mandatory. Miniaturization requires insects to flap their wings at higher frequencies to generate sufficient aerodynamic forces to stay aloft. Adjusting for size, flies might have the most powerful flight muscles on the planet. And their maneuverability is legendary. Despite its outlandishly enlarged eyes, a big-headed fly (one of the Pipunculidae) is able to sustain flight in the close confines of a folded insect net—the volume of a tea bag.

If you have no net, the flight behavior of an amorous male blowfly on the lookout for a passing female is a readily watchable example of flies' aerial prowess. On an April morning while exploring a natural scrubland in south Florida, I encountered one of these males, a greenbottle fly, hovering over a footpath at about eye level. The centimeter-long insect appeared virtually stationary, as if suspended from an invisible string. He seemed almost nonchalant at my presence. Moving slowly, I was able to bring my face to within about one foot of the fly before he edged away just enough to maintain that minimum distance. I slowly reached out until my finger was just four inches from him before he responded. If I raised a hand suddenly, the fly would zoom quickly away, then materialize again two or three seconds later, always in the same spot. At all times he faced the same direction—westward in this case. His wings, beating hundreds of times per second, made a faint, low hum. Several times, he dashed suddenly away when I had made no movement, and I noticed that these dashes usually accompanied the sound of another insect flying by. A few other males hovered in the vicinity, and I suspect my male was chasing away competitors or hoping to intercept a passing female.

Hovering in an open area requires constant adjustments to compensate for tiny gusts and air currents. In the BBC television series *Life in the Undergrowth*, slow-motion cameras are trained on a male hoverfly perfectly lit in a British meadow. We can actually see the fly's blurred wings tilting independently of each other to keep the fly in its spot. Host David Attenborough uses a peashooter to demonstrate the alertness and quickness of the hoverfly. As the pea zooms past, the fly whirls around in an instant and sets off in pursuit. It's an impressive combination of vision and agility—though in this case also temporarily mistaken identity.

The study of fly flight is an active field, with applications to physics, energetics, and robotics. To achieve their state-of-the-art flight abilities, flies use high-tech equipment. Generating frequencies of 100 or more beats per second is beyond the physiological limits of nervous-tissue firing rates. For this reason, the upper limits of fly flight are achieved not by nervous control alone; they are supplemented by mechanical connections.

Flies have evolved a complex system of levers, fulcrums, tiny knobs on wing veins, stretch-activation mechanisms, and a system much like the manual clutch on a car's transmission linked to a sort of gearbox that enables them to control each wing independently. A shieldlike plate, the *scutellum*, connects the two wings, while another plate, the *subepimeral ridge* (a bump on the lower side of a fly's thorax), connects each wing to its respective haltere. A clutch mechanism connecting the scutellum to each wing can be engaged (or disengaged) on either side, decoupling the wings and allowing them to move independently, which enhances maneuverability. The gearbox resides at the base of each wing and is composed of three structures that—like the gearshift of a car—work in concert to adjust the height of the wingbeat from low to high.

Even with all their machinery for lift, flies wouldn't get far without balance and steering. Our system of balance is in our ears. Not so a fly's. Flies balance and steer using their halteres, those remnants of wing pair number two we met earlier. During flight, the halteres resemble beating drumsticks, flapping at the same rate but usually precisely antiphase to the wings. They act as gyroscopes that swing up when the wings flap down, and vice versa. If the fly yaws, rolls, or pitches during flight, the halteres twist at their bases but maintain their original plane of movement. Special nerve cells detect the twists, allowing the fly to correct its orientation.

Their name notwithstanding, some flies lack wings altogether. Their ancestors had them, but—like flightless birds on predator-free islands—their life history trajectories rendered those costly appendages pointless, so over countless generations they lost them. Case in point: bat flies. If you spend your life scuttling about crablike on the bodies of bats, you don't need to go airborne to get from one place to another—the bats do it for you. And deplaning from one host and boarding another is straightforward when hosts huddle together for long periods, as bats do. And so, bat flies, numbering an astonishing 511 known species in two families, have gradually lost their wings over the millennia. I saw a few while studying bats during my graduate student days, and had there not been someone there to explain what they were, I would never have thought they were flies.

If you've wondered about the gravity-defying ability of flies to walk on windows and ceilings, it's thanks to a pair or trio of pads on each foot, called *pulvilli*. From each pad sprouts thousands of tubes, each of which terminates in a very smooth, flat pad. Once thought to work by suction, pulvilli work by adhesion. Tiny drops of a gluelike substance made of sugars and oils oozing through each of those tubes bond them to the smoothest of surfaces using molecular attraction forces. The fly walks by altering the angle of the foot pads to release the hold. House geckos use the same trick to scamper across walls and ceilings as they hunt for insect prey.

The quickness of flies, and the cockiness with which they remain in place or quickly return despite our efforts to shoo them away, is due in part to those bristles and hairs we encountered on our visit with Mark Deyrup. The base of each follicle is innervated, rendering the fly sensitive to minute changes in airflow. This early-warning system helps the fly detect approaching violence, which helps explain why a fly is so hard to swat.

When scientists took a close look at mosquito flight, they discovered something new. Mounting eight slo-mo cameras to film the flight action at a range of angles allowed the creation of a 3-D model of the whining insect's wing movements, which pass through a paltry 40 degrees of motion, barely half that of a bee. This shallow movement shouldn't be enough for a mosquito to fly using only a leading-edge vortex (a trapped pocket of air that helps generate lift). What the cameras revealed was a second vortex at the trailing edge of the wings. As the back edge of the wing chases the path of the front edge, it captures the swirling wake of the previous flap, allowing that energy to be recaptured. This provides the extra lift that permits the mosquito to make a nuisance of herself. That second vortex saves energy by reducing the size of the pathway each wing needs to trace. At 700 beats per second, that amounts to significant savings.

Efficient flight has allowed some flies to evolve impressive migrations, such as those of the marmalade hoverfly, millions of which zoom over the Swiss Alps twice yearly on their round trip from northern to southern Europe. Based on his aerial monitoring of these mass insect migrations, Karl Wotton, an English geneticist at the University of Exeter, extrapolates that many billions of hoverflies of various species migrate across Europe each year—a never-ending stream of tiny bodies glinting in the mountain light. With a tailwind they ride high; with a headwind they scoot low. "They fly fast . . . and they don't stop," Wotton says. "The butterflies are getting turned around like in a tumble dryer, but the hoverflies just shoot straight over."

Motion Sensors

For flying organisms, visual fields change quickly, so—with the exception of echolocating bats—strong vision is crucial. Insect eyes are fundamentally different from ours. Whereas a vertebrate eye is composed of just one unit, an insect's compound eye has numerous separate units, or facets, packed together like the hexagonal chambers of honeycomb. Each facet, or *ommatidium*, is a fully functional organ of sight, which independently sends visual signals to the brain. Insect eye facets are typically about 10 micrometers wide; about 20,000 of them would fit on a pinhead.

This arrangement implies that what an insect sees is a mosaic of small images knitted together. Indeed, that's how my undergraduate entomology textbook presented it, using a schematic diagram to illustrate. It was a blurry, pointillist affair, and it left me musing about the need for a helmet if I had to zip through life with such poor eyesight. But the behavior of many insects—flies included—hints that they have better vision than that, and it is now generally accepted that insect brains integrate the separate signals from each ommatidium into one seamless whole, just as our brain melds the images from our two eyes into one. The insect's compound eye has been the inspiration for research and development of motion-sensing cameras for use in military surveillance.

Flies have teams of neurons working in concert to handle visual challenges at the cellular level. Specialized, motion-sensitive neurons track the optic flow of objects as they move through the fly's field of vision, helping to maintain a flight course. Another set of neurons uses the optic flow to monitor self-motion. A third set of neurons appears to analyze the content of the visual scene itself, such as separating figures from the background by detecting relative flow—a process called *motion parallax*. Three *ocelli*, light-sensitive organs on the top of the head and totally separate from the eyes, detect changes in light intensity, helping the fly to react swiftly to the approach of an object.

Many flies have a more mundane way to deal with the visual

flow caused by rapid flight: they make quick, repeated sideways glances. Flying blowflies, for example, shift their gaze by rapid, discrete turns (*saccades*) of the body and head, keeping their gaze basically fixed between each saccade. (Our own visual systems produce similar saccades when we gaze out through the window of a moving car or train; our eyes briefly lock on to a nearby object, then skip forward to another, creating a rapid side-to-side movement of the eyes.) These rapid movements produce an almost seamless translational optic flow between saccades, enabling the fly to extract information about the spatial layout of its environment. I remember feeling slightly unnerved the first time I noticed a fly suddenly glance to one side. It seemed so purposeful that I half expected it to hail a taxi.

Studies of fruit flies, each of whose eyes has a modest 600 or so facets, have shown that they use a visual prioritizing system. Things that are static are kept blurry, while anything moving, independent of the visual changes caused by the fly's own movements, is brought into sharp focus. As Peter Wohlleben says in his book *The Inner Life of Animals*: "You could say that the tiny tykes are stripping things down to the bare essentials, an ability that you surely would not have expected these little flies to have." We do much the same thing. As you read this you can detect many things on and beyond the page in your peripheral vision, but you are not focusing on those things. Even the words just inches away from the ones you are reading at this moment are blurry. In this way, our vision acts like our conscious minds, which can think about only one thing at any given moment.

Using high-resolution, high-speed digital imaging of fruit flies faced with a looming swatter, Michael Dickinson, professor of bioengineering at the California Institute of Technology, and graduate student Gwyneth Card determined that the insect's tiny brain calculates the location of the impending threat, comes up with an escape plan, and places its legs in an optimal position to jump out of the way. All of this happens within about one tenth of a second after the fly first spots the swatter. In carefully controlled slo-mo filmed experiments using a 14-centimeter (6-inch) black disk (the "swatter"), the curious scientists noted that the flies integrate nearly 360-degree visual information with mechanosensory information from their legs, causing them to lean away from the impending threat by pushing their middle pair of legs toward it. If a fly has a conscious experience of an event like this (see next chapter), then we might add that the notion of escape is accompanied by the emotion of fear.

If you've matched wits hand-to-hand with a fly, you'll know how well their vision serves them by how difficult they are to catch. As a teenager I developed a quite effective technique for catching houseflies with my bare hands while working in a summer-camp kitchen. I would slowly move my hand toward the rear end of a fly resting on a flat surface such as a tabletop or a vertical wooden beam (beware of splinters on naked wooden surfaces!). Once I got within four or five inches of my target, I would pause to ready my nerves for the ambush. Then as fast as I could I would sweep my hand toward the fly and close it with reflexive speed. My target was airborne before my palm arrived, and I had no time to react to its movements. But if my speed was good and my placement adequate, the fly ended up trapped inside my hand. At my peak I probably reached about a 60 percent catch rate, and on rare occasions I even caught two flies in one sweep. A lifelong admirer of flies, I released them outside to a better fate than the sticky fly-strips festooning the ceilings. I could usually feel my captive skitting about inside its fleshy tomb, but not always. Many were the times that I either accidentally released a crafty fly back into the kitchen after I thought I'd missed, or cautiously opened my hand outside to find nothing there.

Different physical characteristics between the sexes usually have something to do with reproduction. True to this tenet, the males of many flies tend to have larger eyes that meet at the midline. These *holoptic* eyes allow virtually 360-degree vision—all the better to find females with. In some extreme examples, such as the big-headed flies, the eyes cover most of the entire head, which looks as if it has been inflated (see the image in the photo insert). Female flies, with few exceptions, have *dichoptic* eyes, which do not meet. I wonder if the more visually endowed males survive more predatory attacks than females, or if perhaps their visual edge is offset by inferior maneuverability.

If holoptic vision has a further drawback, it is that binocular vision might be compromised. Robber flies' eyes are well separated, allowing them binocular vision for good distance perception, crucial in coordinating their attacks on flying prey and probably also in detecting and avoiding an approaching predator. I have found that only by creeping slowly am I able to get within arm's length of a perching robber fly (more on them in chapter 4).

Holoptic or not, good vision doesn't benefit the housefly trapped against a windowpane. A glass barrier must be completely confounding to a visually orienting insect that has no experience with glass in nature. The fly sees only the scenery beyond and is unable to override its compulsion to approach it. As far as I know, nobody has tried to find out whether flies adapt to windows or other manmade phenomena.

Tasters

Flight gets flies from A to B, location B often being a food source. We might think of flies as undiscerning diners, but the equipment they have dedicated to taste suggests otherwise. Like smell, taste involves chemoreception, but it differs from smell in requiring physical contact with a substance. Flies' sense of taste, unlike ours, is not limited to the mouthparts. In addition to their food-sucking proboscis, flies' taste receptors are mounted on bristles scattered all over the body, including legs, wings, and the egg-laying ovipositor. Most notably, the soft footpads of flies contain taste organs. Most humans would, I suspect, not view being able to taste with one's feet as a desirable trait, with the hypothetical exception of traditional grape smashers employed in winemaking. But this ability lets a fly sample the nutritional potential of a ripe banana, an arm, or a tabletop the instant it alights on its surface.

Viewed in extreme close-up, a housefly's spongy proboscis

The proboscis of a housefly is a marvel of structural and functional sophistication.

(© SUSUMU NISHINAGA / SCIENCE SOURCE)

reveals an organ vaguely reminiscent of the prehensile tip of an elephant's trunk. The fly's built-in squeegee mop is lined with a corduroy arrangement of channels through which liquid food is sucked then pumped to the throat. But the mop also has a reverse gear: through those same channels, saliva can be drizzled on the substrate, helping dissolve solid items into a suckable form.

This spongy lower lip of most flies allows them to first liquefy then imbibe sweet or otherwise attractive foodstuffs, such as dried honeydew spattered on a leaf surface by plant-sucking bugs. Fly authority Stephen Marshall suspects flies were feasting on this ubiquitous source of nectar long before flowers evolved and began producing it. Today's flies are 100 times as sensitive to the taste of sugar as we are.

How does the taste actually happen? Careful studies on fruit flies have determined that their taste-sensitive feet achieve their task with tiny hairlike filaments, each of which terminates with a pore. Each pore houses individual neurons sensitive to a different group of chemicals. These and neighboring neurons transmit signals to the fly's brain.

A potential food item must pass two taste tests before a fly will partake. If the encountered substance, such as a smear of marmalade or a puddle of water, passes the foot taster test, then the fly's brain gives the command to extend the proboscis. But the fly does not start to imbibe the substance until it has passed a second test by sensory hairs on the tip of the proboscis. Each of these hairs is hollow with an opening at the tip, inside which lie five cells. Two of these cells are sensitive to salty solutions, and one each to water and sugar. The fifth cell doesn't aid taste; instead it detects surface resistance and resilience via the bending produced when the fly sets its foot down. A careful count of the numbers of taste-sensing hairs on a blowfly (*Phormia regina*) found 308 on the foreleg, 208 on the middle leg, and 107 on the hindleg, plus

250 on the proboscis, alongside 132 chemically sensitive papillae (small, fingerlike projections), for a total of about 1,600 taste sensors per fly.

Despite hundreds of millions of years of evolutionary segregation, flies' sense of taste appears to be similar to ours. Behavioral and genetic studies led by Kristin Scott at the University of California at Berkeley have determined that fruit flies have receptors devoted to sweet and bitter tastes just like humans. And like ours, flies' ability to discriminate tastes is simpler than their odor detectors, which allow for finer discriminations. Another feature of taste behaviors that is conserved from flies to humans is that they are exquisitely attuned to internal state. In Scott's own words: "Animals dynamically adjust the probability of feeding to ensure that caloric consumption and energy expenditure are in balance." Quite simply, a full fly isn't interested in food.

Smelling and Hearing

As we might expect of creatures that can taste with various body parts, flies are good smellers. Flies smell with their antennae. These multipurpose wands are covered with chemoreceptors extremely sensitive to a range of chemical cues, and they respond to scents at much lower concentrations than we do. Some carrion flies can detect a rotting carcass from at least ten miles.

Most of the research on flies' sense of smell has focused around the two main loci of the fly-human nexus: disease-spreading blood feeders and crop pests. Bloodsucking flies home in on chemicals emitted by their food source, and so do flies that feed on plants. Olfactory receptors located on the antennae are specialized for the detection of chemicals characteristic of whatever or whomever they feed on—anything from a turd to a tulip, depending on the type of fly.

There are many chemicals to choose from. Depending on whom you're sampling (and, presumably, when), human scent is made up of 300 to 500 chemical components. A research team at Johns Hopkins University is working to identify the specific components of human scent actively perceived by olfactory centers in the brains of the human-biting mosquito *Aedes aegypti*, an important vector of Zika virus. This fly has three olfactory organs with three families of receptors honed to respond to human odor. The plan is to develop a designer chemical fragrance mimicking human scent that can be used to effectively bait mosquito traps to improve vector control and surveillance to combat Zika and future threats. There is already a device called a Mosquito Magnet, which uses carbon dioxide to attract, trap, and kill mosquitoes.

Flies' versatile antennae have a further sensory use: hearing. Like humans, flies can distinguish different frequencies. The mechanics of fly hearing involve a beautiful cascade of responses, beginning with the detection of air vibrations (sounds) by the distal segments of the antennae, and ending with nerve signals relayed to the brain. The cascade begins with a very, very slight deflection of the antennae—a few 10,000ths of a hair's width. This causes downstream sensory cells to stretch, which opens ion channels through which charged molecules enter, triggering an electrical impulse. At this point, a mechanical amplifier comes into play, a sort of motor that magnifies the effect of the deflections. If the fly is stimulated at a certain frequency, the sensitivity to that frequency is increased with each oscillation, as with a playground swing being pushed. Lower sounds allow greater amplification.

Hearing in flies is usually reserved for courtship, and it's thought that most flies that don't use sound in courtship are deaf. Fruit flies are enthusiastic courters; males use songs generated by rapid wing vibrations to woo potential mates. Studies at the University of Iowa found that fruit flies' hearing deteriorated when they were subjected to loud noise similar to that generated by a rock concert. The assault caused structural damage to nerve cells involved in hearing. As has been observed in humans, the flies' hearing recovered a week later. Long-term exposure to high decibels leads to permanent hearing loss in us, but the much shorter adult life span of a fly probably renders less vulnerable the one who unwittingly finds herself at a Metallica concert.

Masters of Adaptation

With their diversity of adaptations to sense and move through the world, flies just might be the ultimate evolutionary opportunists. As we will see in the chapters ahead, flies have evolved countless ingenious solutions to the various problems of living in a challenging world. The great novelist and humorist Mark Twain found himself full of admiration for a fly that spends much of its time underwater in Mono Lake, California. The tiny alkali fly's waxy, hairy body armor traps air, allowing it to dive to the bottom, where it feasts on algae. Twain delighted in his failure to drown the flies, as he wrote in his travel memoir *Roughing It*: "You can hold them underwater as long as you please—they do not mind it—they are only proud of it. When you let them go, they pop up to the surface as dry as a patent office report."

Alkali flies form their bubble suit by pressing down on the water surface, usually crawling headfirst until the lake forms a dimple. As the dimple grows deeper, the surrounding water pressure reaches a threshold and it suddenly engulfs the fly inside a silvery air pocket. The fly's clawed feet and mouthparts are conveniently free of the air bubble, enabling the insect to scuttle along the bottom. Except for brine shrimp, the fly is about the only organism to live in this highly alkaline lake. "It's a great gig because there's no fish in the lake," says Michael Dickinson, who recently described the fly's remarkable biology more than a century after Twain heaped praise upon it.

Despite increased salinity of the lake since the 1940s, after diversions to Los Angeles of some of the freshwater streams that previously flowed into it, the flies have persisted; their swarms are large enough to attract hordes of gulls that fly through them with their beaks agape. So abundant are these flies that they help support a local ecosystem that attracts about two million birds of more than 300 species that migrate to Mono Lake each spring to feast and breed. Today, new water diversions are further shrinking the lake, raising sodium carbonate concentrations to dangerously high levels, even for the fly. Another threat is that sunscreen from the lake's occasional human swimmers strips the fly's waxy coat, making it more susceptible to drowning.

It would be sad to see them go, but we should note how much more quickly flies can evolve than we can. Little wonder that flies comprise such a treasure trove of physical gifts to thrive on Earth, given that they can squeeze 500 generations into 1 of ours. There is a cautionary lesson here concerning our ability to adapt to man-made change—something we will return to in the final chapter.

Chapter 3

Are You Awake? (Evidence for Insect Minds)

> The genius of life shines more in the construction of smaller than larger masterpieces.
>
> —SANTIAGO RAMÓN Y CAJAL, NEUROSCIENTIST AND NOBEL LAUREATE

There is a joke that asks, "What's the last thing to go through a fly's mind when it hits a windshield?" Answer: "Its butt." Aside from its irreverence, I like this joke for at least implying that a fly has a mind.

Are flies conscious? Do they experience anything? Is a fly a thing, or a being? I think we can safely apply these parallel questions to insects in general. For if any insect is conscious, then I think it is fair to say that any other insect *could be* conscious. But is it really possible for such small creatures to have experiences? At first it may seem unlikely, but if we stop to watch insects closely—to regard their coordinated movements and their complex, seemingly flexible actions—it is harder to imagine them as little blank slates, going about their lives in a mental vacuum, without even a smidgen of awareness. When I regard a fly cleaning his legs by rubbing them together or his wings by swiping them with his hindlegs, or when I watch a wasp or beetle preening her antennae by running them through her mouth, I see a creature with intent. And if you've ever watched a praying mantis swivel her articulating neck to fix her two eyes on your own gaze, then you may have had the eerie feeling of being regarded by a creature who knows you're there.

I'm not claiming outright that flies or all insects are conscious. Indeed, no one can. Trying to ascertain consciousness in another being has been described as "the hard problem" in the life sciences by influential Australian philosopher David Chalmers. But we are not resigned to eternal ignorance. Science has tools to explore the matter, including anatomy, physiology, evolutionary biology, neurology, behavior, and genetics. We also have a powerful emotion, empathy, that helps us take the perspective of another. We can observe other creatures expressing pain, fear, joy, playfulness, anger, etc., and relate their experiences to our own in similar contexts.

Of course, it's one thing to watch a dog chase a ball and imagine that the animal is feeling joy, but it is quite another to ascribe the same feelings to a pair of mating flies. As we venture further from humans on the evolutionary bush, our ability to apply empathy is weakened.

One reason we ought to be cautious in attributing consciousness and sentience where it may not reside is that seemingly intelligent behaviors may occur without conscious awareness. Evolution is a master problem solver. With the luxury of eons of time, and a huge diversity of natural resources to experiment with, evolution has wrought organisms with stunning adaptations. Some of these suggest intelligence that seems improbable.

Here's a dipteran example of such a "clever" adaptation. It involves foresight in the overwintering strategy of a fly called the goldenrod gallfly. In late summer, the adult fly injects an egg into the stem of a goldenrod plant. Chemicals from the egg (or perhaps the mother fly) cause the surrounding plant tissue to form a protective tumor around the developing insect, called a *gall*. This strategy amounts to forcing the host plant to build a custom house with a well-stocked larder of food. By late summer, the hatched larva will have fed on the gall's expanded flesh and grown to its maximum size. At this time, plant, gall, and larva stop growing. Before the first frost, the fly larva then does something remarkably prescient: it burrows to the outermost layer of the gall using its chewing mouthparts, then, just before piercing the gall's surface, retreats back to the gall's center, where it spends the winter. When spring arrives, the metamorphosed adult fly crawls through its premade tunnel, pushes itself through the thin outer membrane, and flies off on its life adventures. The reason the larva bores a channel from the center of the gall to the surface is that, unlike the larva, the adult fly has no chewing mouthparts. By constructing an escape route months ahead of time, the maggot avoids becoming an adult fly helplessly entombed in its winter home.

It seems more plausible to interpret such behavior in a blind fly larva as a product of instinct than as intelligence. At least that's what my intuition tells me.

But the shrewd instincts of a gall maggot do not negate the possibility that insects are aware. Scientific interest in the possibility of insect sentience is growing. In a 2016 paper published in the prestigious *Proceedings of the National Academy of Sciences*, Australian biologist Andrew Barron and philosopher Colin Klein argue that insects might experience things based on the presence of brain features that structurally and/or functionally parallel the brains of vertebrates. For example, *mushroom bodies* support learning and memory, a *central complex* processes spatial information and organizes movement, and an anatomically sophisticated *protocerebrum* connects other brain regions and collects incoming sensory information. The authors conclude that insects might already have been using consciousness to support their active foraging and hunting lifestyles way back in the Cambrian era some 500 million years ago.

In this chapter I present some of the more compelling evidence for conscious awareness in insects, and in flies in particular. I invite you to draw your own conclusions.

A Bowl of Rotting Peaches

Before getting into the science, let me share a personal experience of the sort that may give one pause before concluding, as I suspect most do, that insects are incapable of having conscious experiences.

While visiting a friend's country home in southern Ontario one summer, I noticed a small, white ceramic bowl perched on a kitchen countertop. Nothing unusual about that, I thought, until I peered inside. There I found a most curious sight. The bowl contained several chunks of peach looking well past the threshold of human edibility, and about fifty fruit flies. The bowl was sealed by a film of plastic wrap stretched taut around the rim. Most of the flies were standing around, like guests sipping wine at a cocktail party. Some strolled casually over the fermenting fruit, which was speckled white around the edges with mold. A few others stood or walked across their "glass" ceiling in that gravity-defying way that flies do.

I gazed in wonderment at this strange domestic scene. There is nothing unusual about seeing fruit flies in kitchens. But how in god's name did these end up inside that layer of plastic? Had my friend Celia sneaked up on the bowl with plastic wrap in hand and pounced on it? Such is the wariness and speed of flies that all but a few would surely have taken wing and escaped before the veil came down over them. Were the flies already incubating on the peaches when the plastic ceiling was installed? That couldn't be it, for there were no telltale pupal husks in the bowl.

When she returned from an errand, I asked Celia about the peach bowl, and the riddle was solved. It was a flytrap. The construction was simple enough: place pieces of overripe peach in bowl, seal with plastic wrap, poke a dozen or so tiny holes in the plastic with the tip of a sharp knife, wait a few hours, and voilà: trapped flies.

What?

If you're like me, you're trying to visualize a fruit fly squeezing through a tiny slit in that plastic layer. For one thing, how do they find the holes? That's a matter of what most folks might describe as the alluring aroma of fetid peach seeping through a crack in the wall, and what scientists would call a chemical gradient. The perceptive flies follow the irresistible scent trail to its source. But how do they get inside? How does a little fly squeeze through the slit? I'll come back to that in a moment. The fact is, they get inside, and they have their way with the peach, taking their fill of the succulent liquor and, given enough time, mating and laying eggs.

"It works like a lobster trap," Celia told me. "They can get inside, but they struggle to find their way out."

I was surprised to find the trap in its same spot the next morning, with the same mold-flecked peach looking like the aftermath of a fly orgy. But there were not more flies in the bowl; there were fewer. I grabbed my binoculars (as a birder, I never travel without them, and when you turn them around they make an effective magnifying glass) and came in for a closer look. What I witnessed astonished me. A fly trotted across the plastic ceiling, found one of the slits in the plastic, and proceeded to make its exit. The tiny insect used its two forelegs to pry apart the plastic, squeezed its head through the fissure, then worked away with its remaining

These fruit flies found their way through tiny holes in the plastic wrap stretched over a bowl of fruit and are preparing to find their way out.

(PHOTO BY THE AUTHOR)

four legs to squeeze its plump, peach-filled torso free of the barrier. The operation required a considerable, coordinated effort and took a minute or more to complete. Once out, the fly paused momentarily, then flew off.

"Celia, you might want to take those flies outside now. These traps work in reverse, and some of your flies are escaping back into the kitchen," I told her.

What so captivated me about the flytrap incident was not so much the flies' ability to find their way to the rotting fruit, but the apparent intentionality and determination with which the flies escaped the trap. While it is easy to imagine what drives fruit flies to enter a peach trap, it's puzzling what drives them to want to leave a luxurious source of food and a breeding site. One can argue that instinct "drove" her in, but was it instinct that "drove" her out? It left me struggling to reconcile what I had just seen with a common assumption that flies are nothing more than robotic automatons, with no experience, no awareness.

As I sit in a busy coffee shop recalling the memory of Celia's flytrap, I'm watching a poodle mix snuffling around the café, homing in on dropped crumbs that have escaped under padded chairs, and I'm reminded of how much keener another being's sense of smell can be than mine.

There are other variations of Celia's simple flytrap. In a short online video, Cornell University entomology professor Brian Lazzaro explains that a funnel placed over a jar of wine or overripe fruit will lure fruit flies in and confound their attempts to exit. I wonder if that design, too, becomes less effective with time.

Marla Sokolowski, a professor of genetics and neurology at the University of Toronto and a former teacher of mine, told me of walking into a grocery store where there were lots of flies and telling the manager how to catch them with a trap like Lazzaro's, consisting of a funnel placed in a half-empty bottle of beer (or yeast and water). Marla's daughter, then a preteen, rolled her eyes, embarrassed to have a mom who talks to strangers about flies. When they returned to the store two weeks later, there were noticeably fewer flies and one grateful manager. These traps represent human ingenuity capitalizing on fruit flies' resourcefulness.

A Blurry Line

When it comes to mental abilities, we tend to hold vertebrate animals in higher regard than animals without backbones. We tend to believe that invertebrates lack any mental life whatsoever. But science has been exposing the frailty of such a belief, and a once bold line drawn between vertebrates and invertebrates has become blurry.

The evidence for consciousness in octopuses and their molluscan kin, for example, is fairly unimpeachable. If you doubt this, I encourage you to read *The Soul of an Octopus* by Sy Montgomery, or *Other Minds* by Peter Godfrey-Smith. Octopuses and their close relatives the squids, cuttlefish, and nautiluses (collectively, the cephalopods) have the most complex nervous systems among invertebrates. Octopuses display problem-solving abilities, emotion, play behavior, and unique personalities. They can untie knots, open jars, and gain access to toddler-proof containers. They can learn by watching others, and they are renowned as escape artists. Some experts believe that octopuses were the first creatures on Earth to evolve consciousness, and their evolutionary distance from vertebrates indicates that consciousness has evolved on our planet at least twice.

Moving closer to insects on the tree of life, emerging evidence suggests consciousness might have evolved at least three times. Spiders, for instance, demonstrate intelligent behavior. A notable case is the detour behavior shown by jumping spiders while hunting prey. It was discovered in the 1990s that jumping spiders in the genus *Portia* will back away from prey in search of a more strategic approach where they are less likely to be detected by the prey. These spiders will also move around an object that blocks the spider's view of its prey, which demonstrates "object permanence" by the spider. A more recent study by the same research team showed that 16 species of jumping spider (including 10 from genera other than *Portia*) solved a similar hunting problem that required them to remember a food's location and to ignore a path that led to a nonfood source.

What about cognition in spiders' close arthropod cousins, the insects? There are some compelling findings, especially but not exclusively from the social insects.

By taking close-up photographs of a species of paper wasp,

scientists ascertained that these colonial insects recognize each other by their distinctive faces. In trials in which choosing an unfamiliar face (digitally doctored by rearranging or removing parts, such as the antennae) was punished and choosing a familiar face was not, they chose the familiar faces.

I love the idea that a wasp recognizes the face of a familiar comrade; maybe they also exchange greetings with their antennae. But my favorite study of cognition by insects involves ants. In 2015, Marie-Claire and Roger Cammaerts from the Université Libre de Bruxelles published the first demonstration of mirror self-recognition (MSR) in an invertebrate. The MSR test was first published in 1970. Chimpanzees were anesthetized then marked in a location such as the forehead where they could see the mark only in their reflection. When presented with a mirror, the chimpanzees inspected the mark in the mirror and touched or tried to remove it from themselves. This behavior indicates that the chimps recognized that the image was a reflection of themselves, not another chimp, and it became a benchmark test for self-awareness. Until the ant study 45 years later, only great apes, elephants, dolphins, and magpies had passed the MSR test. (In 2018, a fish called the cleaner wrasse joined the list.)

The Cammaertses studied three species of *Myrmica* ants and found that they behaved differently upon seeing a reflection of themselves in a mirror compared with seeing other colony members through a pane of glass. With the mirror, they behaved uncannily like socialites inspecting themselves before going out for an evening on the town. They rapidly moved their head and antennae to the right and left, touched the mirror, went away from it and stopped, and in some cases groomed their legs and antennae. They also tried to clean off a blue dot placed on the front of their heads, which they otherwise ignored if they could not see themselves or if the blue dot was applied to the back of the head where it was not visible in the mirror. Brown dots, which matched the ants' body color and were thus camouflaged, likewise were ignored. Aware of the potential for their study to ruffle feathers (and fur) in the scientific community, the Cammaertses hastened to add that their findings need not necessary imply self-awareness by ants.

Isn't it interesting how readily we acknowledge self-awareness in a mammal but grasp for alternative explanations in an insect because it defies our biased expectations? In an earlier book (*What a Fish Knows*), I cited many scientific studies that belie the common prejudice that places fishes below the other vertebrates, especially mammals and birds. It's a much sterner challenge for insects, and I'm not claiming equality here, but it is a recurring pattern that when we peer closer, animals yield new surprises. It reminds me of the famous words of Louis Leakey when he received news of tool use by chimpanzees from his prodigy Jane Goodall: "Now we must redefine tool, redefine Man, or accept chimpanzees as humans." What certain insects are now known to be capable of casts serious doubt on a lot of our currently held cultural biases against them.

Ants and other insects also use tools. Funnel ants use bits of leaf, wood, or mud as sponges. Holding the sponge by her mouth, the ant dangles it into a desired source of nourishing liquid (e.g., fruit pulp or the body fluids of prey) before carrying the wetted sponge back to the nest. This technique enables an ant to transport ten times more fluid than she could otherwise carry. A New World ant of arid desert regions surrounds the colonies of competing ants then drops small pebbles and other debris down the entrance holes, giving the marauders more time to forage undisturbed. The use of leaves by leaf-cutter ants on which to grow fungus not only qualifies as tool use but also farming. Digger wasps use flat pebbles as tools to tamp down the dirt, disguising the entrance to the burrows in which they have buried paralyzed prey to provision their egg once it hatches. One type of assassin bug—a predatory insect with beaklike sucking mouthparts—uses the husk of a sucked-dry termite as bait to lure and catch other termites. The assassin jiggles the dead termite outside the termite nest entrance hole, then grasps a new termite that comes to pull her comrade back into the nest. If the bug successfully grabs its new victim, the previous one is immediately dropped, and the process repeats itself. One bug caught and chugged 31 termites this way before waddling off with a very distended stomach.

Prevailing scientific opinion has been that these are the mindless mechanical workings of instinct bereft of conscious experience. But there are reasons to be cautious in drawing such conclusions. A more in-depth study of tool use by funnel ants conducted in 2017 found flexibility in the ants' choice of tools for transporting liquids back to the nest. The ants learned to favor superior manmade tools (sponges) for the purpose, and they sometimes modified those tools—breaking the provided sponges into smaller pieces—to enhance their utility.

The strongest challenges to notions that insects cannot think come from honeybees, which have been the subject of a large body of research since Nobel Prize–winning Austrian biologist Karl von Frisch discovered their now-famous "waggle-dance" language in the mid-20th century. In addition to their mind-blowing facility with multisensory symbolic communication to share the location of distant food sources, bees have accumulated an impressive résumé of mental skills. They can recognize human faces. They understand the concepts of "same" and "different" and can transfer these concepts across different visual modes (shapes to colors) and even across sensory modes (shapes to scents). Bees also seem to understand the concept of zero: when trained (with a sweet reward) to fly to images of fewer dots or symbols (rewarded for picking, say, 3 dots and not 5), they tended to favor a blank image (zero) over an image with a single dot.

Honeybees also appear to have metacognition—that is, knowledge of their own knowledge. Taught to fly to rewarding targets contingent on the target's size, shape, and color, bees were more likely to opt out of tough discrimination tasks when there was a bitter taste penalty for getting it wrong. "This suggests that the bees were only taking the test when they were confident of getting it right," said Dr. Andrew Barron, a biologist at Macquarie University and coauthor of the study.

Fly Minds?

Most of the research on flies' inner lives has focused on fruit flies. This is not because fruit flies are the Einsteins among flies, but because they happen to be one of the most-studied animals on Earth. Fruit flies have the advantages of being easily and cheaply bred and maintained in captivity, and having a two-week generation span that lends itself well to genetic studies, as we'll see in chapter 9. Once again, we must use caution in assuming that mental achievements of one fly species apply to other flies, but what fruit flies can do nevertheless indicates what other flies *might* be capable of.

Although human and fly brains differ dramatically in size—100 billion neurons compared with a fruit fly's 135,000 (see the image in the photo insert)—there are some organizational similarities. For example, the fly brain is, like ours, largely divided across its midline, and the molecules and processes driving our brains and theirs are similar. And dopamine and serotonin control arousal in both flies and humans. Like ours, fly brains manage spatial representation, a critical ability for an animal that flies. In the fruit fly this ability resides in a brain region called the central complex, whose functional equivalent in mammalian brains is the superior colliculus.

As we've already seen, fruit flies are resourceful little creatures that can solve problems, certainly when it comes to squeezing through a tiny plastic portal. What else are they doing with their brains?

Fruit flies can easily be trained to associate an odor with an electric shock, and they exhibit short-term, intermediate-term, and long-term memory of these experiences when later tested on the bad odor and another odor paired with no shock. These memories persist after the fly awakens from general anesthesia, and when new nerve cells have replaced old ones. Fruit flies also have an attention span, showing anticipation of a repeating visual stimulus (a black symbol painted on the inside of a rotating drum inside which a tethered fly is flying), waning interest when the stimulus repeats monotonously, and renewed attention when it changes (e.g., the first symbol is replaced by a different one). Another hallmark of attention is the tendency to suppress and ignore competing stimuli; a fly is less likely to notice, say, another fly nearby while she is fixated on a new symbol in the drum.

And they sleep. When scientists at Washington University School of Medicine in St. Louis looked in on their captive colony of fruit flies one morning, it appeared they had all died, but when the researchers tapped on the glass of the container, the flies gradually roused. It had been nap time. When evolutionary biologist Bruno van Swinderen, at the University of Queensland, recorded fruit flies' brain activity and responsiveness to mechanical stimuli, he discovered that, like us, flies enter lighter and deeper stages of sleep. Their need for sleep rises if they are deprived of it, and if the flies' brains are taxed with learning activities during the day, they need deeper sleep that night.

While they're awake, fruit flies exhibit rational decision making. Observing 2,700 fruit fly matings, researchers at the University of British Columbia found that males were remarkably adept at choosing the female mate who would produce the most offspring. They could do this when there were as many as ten potential females to choose from. Analysis of their large data set shows that the flies use transitive rationality; that is to say, if A is greater than B and B is greater than C, they know that A is greater than C.

A conscious brain should show heightened nerve activity when the animal is engaged in some way. If flies have active minds, might we be able to see their brains in action? To explore the possibility, researchers at the University of California at San Diego performed head surgery on some male fruit flies. They removed a tiny piece of the fly's exoskeleton from the top of the anesthetized fly's head and glued on a tiny see-through panel. Then, allowing the fly a day of recovery, the team tethered the fly to a fine thread and used a laser and a three-camera setup that rotates with the fly's body movements to track electrical activity in its brain while it was courting. Whereas tethered flies wouldn't (couldn't?) court females, untethered flies did so. The brains of noncourting flies remained almost completely dark, while those of courting flies lit up red, yellow, blue, and white. This study doesn't let us experience what the fly experiences, but it shows that an active fly's brain is engaged in a coordinated fashion. That, to me, looks rather like consciousness.

I don't wish to imply that male fruit flies do all of the wooing and choosing. Another study of mate choice demonstrated observational learning by the ladies. When female flies could watch artificially tinted males attempting to mate with another female, they chose males according to the males' successes or failures. For instance, if a green male approached and mated successfully with a female, and a pink male failed to hook up with another female (known by the experimenters to be unreceptive), then when the observer female was presented later with a green and a pink male, she chose the green male as her mate. Switch the color assignments, and the female favored a pink male. Females who could not observe directly the mating outcomes of tinted males showed no such discrimination. In another experiment, females were swayed by their peers, choosing males in physically poor condition over their healthier brethren after seeing model females consorting with the less fit males. This result shows that fruit flies may be influenced more by social factors than by their own judgments. This "mate choice copying," in which the perceived attractiveness of mates is affected by the opinions of others, is widespread in the animal kingdom, including in human women. "I'll have what she's having!"

Often with good reason, scientists have tended to steer clear of anthropomorphism, the attribution of human qualities to nonhuman animals. Nevertheless, the American ethologist Donald Griffin (1915–2003), in his groundbreaking books about animals' inner lives, urged us to be no less cautious when making anthropo*centric* comparisons between humans and insects. "How can we be certain about the critical [brain] size necessary for conscious thinking?" asked Griffin in his 1981 book *The Question of Animal Awareness*. A fly's neuron quota might be a fraction of our own, but 100,000 or more is still a lot to work with. In fact, 100,000 neurons have vastly more potential connections with each other than there are grains of sand on Earth. As we've already seen, insects do some pretty neat things. Furthermore, even if insects and vertebrates don't have a shared conscious ancestor, a useful attribute like consciousness may evolve more than once. If an octopus can do it, why not an insect?

Fly Pain?

If flies have conscious minds, can they feel pain?

The subject of pain has special importance due to pain's unpleasantness and the urgency with which an animal experiencing pain wants to escape it. It is these qualities of pain that give it so much moral heft, for if a creature can feel pain, then it may suffer. We must be careful, however, to distinguish pain, as a felt experience, from *nociception*, which refers to a purely mechanical reaction to a *noxious* (both words from Latin *nocere*, to harm) stimulus without any negative sensation. Without consciousness, even the most neurologically complex body experiences no feelings, no pain, no suffering—and we can be grateful for the discovery of general anesthesia for that.

Opinions vary on whether insects can feel pain. In 1984, Australian scientists felt that available evidence did not support pain in insects, at least not as it occurs in humans. Even so, they recommended anesthetizing insects as a desirable practice to guard against the possibility of pain and to preserve an attitude of respect toward living organisms "whose physiology, though different, and perhaps simpler than our own, is as yet far from completely understood." The eminent insect physiologist Vincent Wigglesworth believed that insects experience visceral pain as well as pain caused by heat and electric shock, while damage to the exoskeleton apparently causes no pain. Insects don't limp on injured limbs (unless a limb is wholly or partly missing, which necessitates a mechanical "limp"), and they don't protect a damaged leg the way an octopus may do an injured arm. In a careful and critical review of physiological and behavioral methodologies, another British biologist, Marian Dawkins, in 1980 also concluded that insects have some capacity for pain. Evolutionarily, an awareness of pain is such an enormously adaptive mechanism that it is unreasonable to simply assume it's unique to vertebrates. In Dawkins's words: "Pain may be expected in organisms whose survival can be augmented by the experience of pain either as part of an escape mechanism or as a basis for the capacity to learn from past experience." Insects need to escape things, and as we've already seen, they can learn.

How do scientists study pain in insects? Here's a modern laboratory setup that uses painful stimuli to study conditioning in fruit flies. The fly is suspended by the thorax—usually with a tiny dab of warm wax or glue—in the middle of a circular arena like the earlier one used in studies of attention, on whose walls can be presented various visual stimuli. By pairing a particular stimulus—say, two vertical stripes—with an aversive outcome (in this case, a beam of unpleasant heat), the fly soon learns to avoid the stripes by flying away from them. The rig is designed so that merely turning away from the stripes triggers the heat source to switch off. This way the fly controls the environment. In another protocol, called the heat box, a fruit fly must learn to avoid the half of a small dark chamber that is heated every time the fly walks in. Leaving the punishing half triggers the chamber temperature to revert to normal. Flies soon learn to keep to the safe half of the chamber, and they will continue remembering to do so if the fly is removed from the chamber and then tested two hours later.

How do insects respond to pain-relieving drugs? Praying mantises, crickets, and honeybees respond to injections of the pain-killing opioid drug morphine with lower defensive responses to an unpleasant event, and the strength of the response is proportional to the morphine dose. This analgesic effect can be blocked by a drug, naloxone, that counteracts morphine's effects in vertebrates. These studies suggest that insects have a general sensitivity to opioids similar to that of vertebrates.

It has been known for decades that rats suffering from arthritis, and lame chickens, choose to drink water spiked with pain-suppressing analgesics, whereas uninjured animals prefer unadulterated water, but it wasn't until 2017 that researchers decided to see if ailing insects will self-administer a pain-relieving drug. Three scientists at the University of Queensland gave experimentally injured bees the option of drinking sucrose water with or without the addition of morphine. The bees were subjected either to continuous pinching of one leg with an attached clip, or to amputation of one distal leg segment. The results were uninspiring. Bees in either group showed no preference for morphine, but amputated bees drank twice as much of both solutions as did uninjured controls. The team tentatively concluded that, while the study provides no evidence that morphine is effective in reducing possible pain in bees, the bees may be increasing nutrient intake in response to the increased energetic demands of healing a wound.

Fruit flies avoid other sources of pain. Trained to expect a mild electric shock following exposure to a chemical odor, they soon learned to avoid the odor. When the training was reversed so that the shock preceded the odor, the flies approached the odor, apparently associating it with relief of waning pain following a shock. Fruit flies also exhibit *second-order conditioning*: trained to avoid an odor paired with an electric shock, they learn to avoid a second odor paired only with the first odor.

Fly larvae might also be sensitive to painful events. Marla Sokolowski has found that fruit fly larvae have painlike escape responses to attacks from parasitoid wasps who try to inject them with an egg through their sharp ovipositor: they curl up, and/or they perform a rolling movement. That they respond similarly to a heated probe shows that the behavior generalizes across at least two different types of pain: mechanical (the pierce of the wasp's egg needle) and heat.

Clearly, there is much yet to be resolved on the matter of whether or not insects feel pain. The evidence so far hints that insects are sentient, but that the location of and expression of their pain can differ from ours. The experience of pain, or not, might hinge not just on a physical event, but also on context. Death by drowning sounds horrible to me, but it is part of the life history of every successful mayfly, whose eggs and larvae can develop only in a lake or pond. That gives me pause when I think about ascribing to the mayfly the pain and suffering we associate with drowning. Perhaps succumbing to water actually feels good (or not bad?) for a mayfly in the throes of reproduction.

Other Feelings

If an animal can experience pain, it probably also knows pleasure. Once again there are some intriguing parallels between insects and vertebrates in how pleasure might be generated. Different regions of insect brains are connected in interacting circuits, with sensitivities to octopamine and dopamine—compounds linked in vertebrates to pleasant (and in some cases unpleasant) feelings.

Then there is behavior. What we currently know about insects' responses to rewards is limited almost entirely to research on honeybees and fruit flies. With bees, the typical method is to see if the bee sticks her tongue out toward a stimulus that has been associated with a sugar reward. With flies, the usual apparatus is a T-shaped maze with a different smell presented at each end, one of which is linked to a sugar reward. The sensory system used with bees is taste; with flies, smell. Once again, we find tantalizing parallels between insect and vertebrate brains.

Studying feelings in insects is made more difficult than studying them in mammals by the fact that their rigid, expressionless head capsules don't show observable facial expressions. But that doesn't leave us in the dark. One of the clues to the experience of something is the degree of response. For example, hungry fruit flies perform better than satiated flies in a learning task rewarded with food, presumably because they want to gain the reward more. This suggests motivation.

The motivational effects of hunger also influence a fly's fear response. When fruit flies at varying degrees of satiation and hunger were given access to food but then subjected to a visual "threat" in the form of a shadow that passed overhead (produced by a rotating paddle between the light source and the flies), the flies showed various defensive behaviors, including running, jumping, or freezing. How fast and how often the flies performed these behaviors, and how long it took for the dispersed flies to return to the food, all showed the important quality of scalability—they increased with the number and frequency of the shadows. It appears that the flies were truly spooked by the shadows and that they made the flies cautious. Consistent with an emotional response to food, hungrier flies (starved for a day) were harder to scare off. Lastly, when subjected to a stressful situation from which they cannot escape, flies exhibit a "learned helplessness" response well known from studies of rodents: they give up. The authors of this study—a team of American researchers led by William T. Gibson of the California Institute of Technology—stop short of calling the fruit flies' experiences emotions, preferring instead to call them "emotion primitives" analogous to fear.

Given the audacity of some flies' predatory and parasitic habits, one has to wonder whether being able to feel fear would always be adaptive. It may depend on the type of fly and the context. Consider a female mosquito, whose mission is to approach, harpoon, and remove blood from a large, sentient, alert mammal with a swatting tail or slapping hands. That has to be one of the most hazardous job descriptions in nature. If mosquitoes harbored too much fear of getting squashed, a lot more of them might starve. But some fear and caution may serve them well as they seek to catch their target unawares, and the skittish habits of stalking mosquitoes and horseflies fit the mold.

Personalities

Animals capable of courting, learning, and feeling fear might be expected to show personalities. Individual variation is as fundamental to evolution as grapes are to wine. After all, how could natural selection operate if there were nothing to select from? That said, we may not expect to find sophisticated individuality in relatively simple animals or in plants. The idea that an amoeba or a tapeworm might have a personality seems far-fetched, but flies are physically and behaviorally more sophisticated.

To test for personality in fruit flies, a research team from the Rowland Institute, at Harvard University, developed an automated device called FlyVac, which measures phototaxis—a response either away from or toward light—of several individual flies simultaneously. Even though a particular fly's choice (light or dark) in one trial was a poor predictor of its choice in a successive trial, idiosyncratic patterns of light/dark preference emerged over the course of the 40 back-to-back trials each fly got. To the team's surprise, personalities varied among individuals within a range of genetic strains, including virtually identical fly strains (achieved by inbreeding) reared under identical conditions. Phototaxis in fruit flies was found to be not genetically based, and it persisted for the fly's adult lifetime (about four weeks). As if to rub it in, inbred flies actually showed higher variability than did more genetically variable strains.

The impetus for having an automated device to run flies through this assay can be appreciated from the fact that 17,600 flies were used in the study. I can't say I envy them their time in the FlyVac, which, after each trial, uses "pulses [to] whisk the fly back into the start tube of the T-maze (initiating a new trial), where an injury-mitigating 'vacuum trap' catches the fly on a cushion of air." (I don't know about you, but I'd be leery about entering any apparatus featuring an "injury-mitigating vacuum trap.") It's a wonder the flies continued to walk forward at all, and indeed they became more reluctant as the trials proceeded.

If flies show personality in their attraction or aversion to light, might they show it in response to, say, smells or landmarks? That hasn't been tested yet, but at the very least what we can conclude from the FlyVac study is that personality traits penetrate deeply throughout animal life. The study authors note that "we even observed this phenomenon in insects of a different order: wild caught white clover weevils, strongly suggesting that behavioral idiosyncrasy [personality?] is ubiquitous."

Some of you might find it a bit of a stretch to interpret individual differences in attraction to light as being akin to personality. So do I. It seems to me that personality requires a constellation of individual differences in different but related situations. However, as interest in describing personality traits in different animals grows, insects might surprise us. It wouldn't be the first time.

What are the implications if insects are sentient? The history of our understanding of animals has been marked, and marred, by underestimation. Ever since Aristotle proclaimed in the fourth century BCE what was most likely already a widespread assumption: that humans are superior to the other creatures, we have held ourselves in higher esteem than all the rest of creation (though inferior to God and the angels). This conceit was amplified in the 17th century when the influential French philosopher René Descartes reasoned that nonhuman animals—assumed to be lacking thoughts, feelings, and souls—were nothing more than complex machines and therefore ineligible for our moral concern.

Two centuries later still, history took a significant turn. Charles Darwin's contributions to our understanding of evolution demonstrated animals' shared biological kinship with us, debunking the Cartesian divide and setting the stage for more informed and enlightened times. It was almost another century, however, before the study of animals' inner lives became widely accepted in science. Today, we enjoy an unprecedented openness to investigate animal minds and emotions, now including the inner lives of insects.

It remains to be seen how our opinions about the inner lives of insects will evolve in the future. The more I think about how animals experience their lives, especially animals only distantly related to my own species, like octopuses and insects, the more I feel that we should be wary of using the human as a template for them. The rich diversity of life-forms on Earth shows us that there are many different pathways evolution can take. What is painful to one creature may not be so to another, and vice versa, but we might give them the benefit of the doubt when there is reasonable doubt. Philosophers call this the precautionary principle.

The best adjudicator might be simple common sense. While reading in the outdoor seating area of a neighborhood coffee shop in south Florida, a tiny insect crawled along the edge of the page. It was too small for me to tell what kind of bug it was, but as it was carrying a piece of luggage larger than itself, I surmised it was an ant. I couldn't coax it onto a leaf, but it readily boarded my finger. When I dislodged it onto a plate, a chunk of the baggage dropped off. If this tiny creature was driven only by instinct, then I would expect it to have continued on its way according to a "have load will carry to destination" algorithm. Instead, the tiny beast walked over to the errant fragment, hoisted it onto its back, then continued on its way. I don't know about you, but I am hard-pressed to imagine flexible behavior like that coming from a creature without a shred of consciousness.

Part II

How Flies Live

Chapter 4

Parasites and Predators

> Great flies have little flies upon their backs to bite 'em,
> And little flies have lesser flies, and so ad infinitum.
> And the great flies themselves, in turn, have greater flies to go on;
> While these again have greater still, and greater still, and so on.
>
> —AUGUSTUS DE MORGAN*

Now that we know what flies are, how they're built, and why we think they may lead conscious lives, the next five chapters explore how flies make a living. Be prepared for some strange, far-fetched, and lurid lifestyles.

Flies are accomplished parasites, practicing a survival strategy that involves spending at least part of one's life history living off the tissues of a host organism, usually without killing the host. Flies are also major practitioners of the more sinister, *parasitoid* life strategy. A parasitoid is like a parasite on steroids. The perpetrator invades the body as a parasite would, but instead of keeping the host alive, the parasitoid ultimately kills the host, having first plundered it for resources. All of the 10,000 or so named species of flies in the family Tachinidae, informally dubbed killer flies,

*De Morgan's rhyme was inspired by Jonathan Swift's satirical poem "On Poetry: A Rhapsody," from 1733. The original stanza features fleas, but flies suffice, if you please.

are parasitoids, all of which develop inside the body of another insect. Because most of those insect hosts are plant-feeders, the parasitoids are enormously beneficial in controlling agricultural "pests."

Parasites and parasitoids have an uncanny ability to feed first only on the nonvital tissues of their hosts. For obvious reasons this is a useful strategy; killing one's food supply too soon means the killer will soon go hungry.

Case in point: snail-killing flies, more commonly known as marsh flies, whose predatory larvae will burrow into snail egg masses to engulf and consume eggs. Adult marsh flies attack snails by biting the foot—the fleshy part that glides along the ground. The snail responds by withdrawing its foot, which conveniently conveys the fly larva into its shell. In some larval species, a single snail is all that's needed, but others kill several snails sequentially. Inside its snail host, the grub feeds strategically to sustain a fresh larder, ensuring that the host snail stays alive for the several days it takes for them to eat their fill. In keeping with such "restraint," some types of marsh fly larvae feed compatibly in the same snail host, refraining from cannibalism. In other species, however, only one marsh fly larva ever emerges from a snail, presumably because the first larva to enter the host eats all newcomers thereafter.

A particular form of parasitism, kleptoparasitism, refers to the habit of taking food from a host. For example, adults of at least one type of gall midge sip body fluids from prey caught by spiders. Some kleptoparasites are highly particular about their food source; certain neotropical biting midges are attracted only to termites captured by one kind of Amazonian comb-footed spider, which suspends its prey in balls hung by silken threads. These dangling morsels are eaten buffet-style by the interloping midges, aptly named freeloader flies (see the image in the photo insert).

Kleptoparasitic termite flies appear to use mimicry to mollify their hosts. Upon entering a termite nest, the adult female flies undergo a strange transformation. They shed their wings, then their abdomens swell into an effigy of a pale termite sprawled on the fly's back. The hoodwinked termites accept their impostors, who help themselves to the termites' food supplies.

Certain other fly parasites achieve their deception more by behavior than by appearances. Mosquitoes in the genus *Malaya* practice a remarkable bit of kleptoparasitic manipulation with ants that have been feeding on sugars. The mosquito performs a wobbling dance while hovering in front of the ant. The fly then extends its forelegs and strokes the entranced ant's head. The caressing and the dancing, with maybe a whine accompaniment from the mosquito, induce the ant to open her mouth, from which the mosquito sucks up the sweet slurry through its strawlike proboscis. If you think that sounds gross, then perhaps you haven't considered where honey comes from.

You might guess that such an abundant and diverse food reservoir as flies would also play host to their own parasites. Among the most manipulative of these is a *Cordyceps* fungus whose spores, upon infiltrating a housefly, sprout tendrils that suck up nutrients, causing the fly's abdomen to swell. When the fungus is sufficiently mature, it turns the fly into a fungus-serving robot. The fly gets an irresistible urge to go to a high place, such as a shrub or a building eave. Arriving at its perch, the fly sticks out its feeding proboscis and uses it as a clamp, gluing itself to the perch. There affixed, the fly buzzes its wings a few minutes, then locks them upright while its abdomen is aimed skyward. The fly dies in this position as the tips of the fungal tendrils push their way out of the insect's body. The fly corpse has become an ideal rocket launcher for the spores, which catapult upward from the tendril tips, creating an airborne mist whose spores are more likely to land on other, unsuspecting flies, to start the life cycle again. Even the timing of the spore ejection is under the fungus's control. Spore release is delayed until sunset, when cool, dewy air provides the best chance for spores to develop quickly on their fly hosts.

Getting under Your Skin

There's death, and then there's *myiasis*. With rare exceptions, myiasis is not a word one wants to know from personal experience. It is the condition of having a fly burrow under your skin. It is not the adult fly who does such an audacious thing, but rather the maggot, which enters after hatching from an egg that the adult fly has deposited on or near your warm, inviting body. The story that opens this book is an example of myiasis, but the creature that invaded my tissues was a puny flesh fly. Grander variations on the theme can be found in nature.

Adult botflies, while often impressively large, are short-lived and furtive, and not often seen by us. In the 60 years I've trodden the Earth as a curious naturalist, I have only once encountered an adult botfly. It was at my childhood summer camp in southern Ontario. A large, bumbling fly of handsome black and white attire, and notably no sign of any sort of a mouth, showed up on a mown knoll near the dining hall. I think it might have recently emerged from its pupa, for it was easily held in my hand while I admired it for about 30 minutes before I released it. It was some years later, when I happened upon a photograph of a similar fly in a book, that I came to know what it was.

At least one botfly species, the human botfly (*Dermatobia hominis*)—which ranges from southeast Mexico through Central and most of South America—has devised a means of avoiding contact with its hosts that is as improbable as it is ingenious. *D. hominis* deploys a mosquito or some other biting fly to do the dangerous work. An egg-laden mother botfly finds and catches a female mosquito and glues an egg onto the body of her much smaller dipteran cousin. Then she releases the mosquito. If that mosquito successfully finds a blood meal, that's good news for the botfly. The damp heat of the skin of the mosquito's quarry stimulates the botfly's egg to hatch, and the tiny grub either drops or quickly clambers down onto the unsuspecting host. The mosquito leaves a convenient hole through which the bot grub squeezes to gain access to the host's interior. There the host will provide room and board to the developing maggot for the next six weeks or so. Upon reaching the size of a small olive, the maggot squeezes its way out through the breathing hole it has made in the host's skin, whence it drops onto the ground to pupate. In one fell swoop, the poor host creature (it could be you) has become a blood donor and a future flesh purveyor. If that mosquito happens to be harboring a microbial disease, such as malaria or dengue, then the poor mammalian host might be the victim of a parasite hat-trick.

Over 40 species of mosquitoes and other flies, and one tick, are known to have been used as human-botfly egg couriers. And despite their name, human botflies do not target only humans. When I contacted Thomas Pape, a curator and botfly expert with the Natural History Museum of Denmark, he explained that these botflies are restricted to hot and humid forests in the New World, and that cattle probably make up the most important host for *D. hominis*, dogs also being frequent hosts.

"There is a nice story here," Pape continued, "and we don't yet have the full answer. Quite possibly, the original host of *D. hominis* was one or more of the megafauna species that became extinct with the arrival of humans some 11,000 years ago. As the native hosts disappeared, *D. hominis* survived in humans and dogs. Cows and other livestock came much later. The original host could very well have been an elephant, of which several species roamed the Americas in prehistoric times. However, this is very difficult to prove as we have no mummified elephants in the New World that could leave traces of such botfly infestations." But we have for the Old World. The mammoth botfly is known from a single larva found dwelling in the stomach of a 100,000-year-old woolly mammoth exhumed from Siberian permafrost in 1973.

Some botflies take a more direct, if no less audacious, approach to their hosts. Snot bots are so named for their devious skill at squirting a jet of larvae, on the wing, directly into the nostrils of their hoofed hosts, typically sheep or goats. Once mature, the larvae make their way to the host's sinuses, whence they are sneezed or snorted out to pupate in the ground. You have probably noticed by now that flies are willing to sacrifice aesthetics for personal gain.

Despite their strange and diabolical habits, the botfly lifestyle has worked very well for them. There are snot bots that develop in the throats and sinus cavities of deer, reindeer, moose, and caribou. There is a sheep-nose bot, a horse-nose bot, and a camel-nose bot. Elephants, gazelles, antelope, and warthogs are among their African targets. Host specificity may reduce competition with other flies, but in the modern Anthropocene it has drawbacks. Owing to perilous declines in its host populations, the rhinoceros-stomach botfly is one of Africa's rarest insects. It also happens to be one of Africa's bulkiest flies, reaching nearly two inches. If the black and white rhinos disappear altogether, the likely extinction of their botfly parasites would not be without precedent.

Robert's Maggot

If you're like me, there's a morbid corner of your brain curious to know what it's like to have a botfly feeding on your flesh. A colleague tipped me off to Rob Voss, a curator of mammals at the American Museum of Natural History (AMNH) who had crossed paths with a botfly in French Guiana. So I contacted him. Rob confirmed the encounter and shared an account of it with me. It began during a six-mile hike, from a rain-forest field site at Les Eaux Claires, where Rob and his wife and colleague, Nancy Simmons, had spent a week netting bats, to the town of Saül, in central French Guiana.

"The day was warm," Rob told me, "and I exercised the male privilege of removing my shirt as we walked. This explains, perhaps, why the egg-bearing mosquitos alighted on my back, a broad target, rather than on my scalp, legs, or arms, the more usual places to become infested. It also may explain why I was parasitized and my companions, who kept their shirts on, were not." (Rob and Nancy were accompanied by a graduate-student field assistant.)

Rob remained completely unaware of his tiny guests until several days later, back in suburban New Jersey, when he began experiencing "odd little pricking sensations." An examination by Nancy yielded three tiny, innocuous-looking red spots resembling irritated mosquito bites. Within a few days, the pricking sensations were becoming somewhat more uncomfortable, and the red spots were now visible as small lesions. Still clueless, Nancy applied hot compresses, which might have caused one of the still-unseen larvae to die. However, the other two sores continued to grow, and the periodic sensations became somewhat painful, so Rob decided to go to a dermatologist in the city.

By now you may be wondering: Why on earth didn't Rob—a biologist with knowledge of the tropics—suspect he had botflies? He figures there are two reasons:

"One is that I had never had bots before, nor did I know anyone who had. (Local people recognize the early stages of an infestation and squeeze the larvae out before they dig deep enough for that to be impossible.) The second is that I could not see the wounds myself, and my wife described them to me as open sores, so I thought I might have dermal leishmaniasis, a protozoan disease that progresses slowly and is rather easily cleared up, so I had no sense of urgency."

The dermatologist examined the sores, noting their perfectly round shape and clean margins. Proclaiming that he had never seen anything like them, he promptly recommended a "tropical specialist," whom we will call Dr. X. Rob, still working through a backlog of tasks that had accumulated while he'd been in the field, waited a couple of days before making an appointment with Dr. X.

"Dr. X seemed eager to see me," Rob reported, "presumably having heard from the dermatologist. After a brief exam and a small, sharp pain that I now believe to have been a lidocaine injection, he called for a nurse. 'I think you have a myiasis,' he said. 'Let me just make sure.'"

Things happened literally behind Rob's back that he could not see or feel.

"'Ah, yes, here it is!' Using forceps, he presented me with a large conical chunk of bloody flesh with a small fly larva wriggling at the apex.

"Now, I am not in the least squeamish," Rob told me, "nor am I reluctant to submit to necessary surgical procedures, but there was a distinct lack of informed consent to this procedure. I was never asked whether it was OK to cut a big divot of tissue out of my back, nor was there any discussion of treatment options. Instead, I suspect that Dr. X just wanted a specimen for his collection. When he announced, 'Now let's get the other one,' I said no."

Before visiting Dr. X, Rob had begun to suspect he might have a botfly infestation, and he'd done some reading. He learned that botfly lesions never become infected and usually heal cleanly. He also knew of at least two biologists who had raised botfly maggots to pupation. Motivated by a scientist's curiosity and miffed at Dr. X for having taken *his* parasite without asking, Rob informed Dr. X that he was going to keep his last larva, thank you very much.

"He squawked with indignation," said Rob, "and tried to talk me out of it, but my mind was made up. Science would triumph over fear."

Dr. X grumpily sutured the incision.

"When I asked for my larva back, he offered to waive the price of my procedure in return for keeping it. I still regret that I accepted his offer."

Returning home, Rob shared his decision with Nancy, and—herself an accomplished mammalogist working in the same department as her husband—she was immediately supportive. And so began what they dubbed "The Botfly Watch." What they did not know (among many other things) was that these larvae normally take about eight weeks to mature, and only two weeks had passed since the couple had returned from French Guiana. So there were six long weeks of parasitism left to go, during which time the larva would grow much, much larger.

"My guest and I settled into a routine," Rob told me. "Most of the time the larva was quiescent and I was hardly aware it was there, but every few hours it seemed to need to shift position, tossing and turning in its fleshy bed, so to speak, and this was briefly uncomfortable. I think the sensation was caused by the rings of tiny hooks that botfly larvae use to anchor themselves in their burrow, and by which they probably move up and down in it. Much less pleasant were periodic evacuations of brownish ammoniacal fluid, metabolic waste products, I assume, which stung like crazy. This usually happened once in the early afternoon and once at night, usually in the hours just after midnight."

As the larva got larger, so did its breathing tube. Most of the time the humans kept the burrow orifice loosely covered with a gauze bandage to keep the excreted fluid from staining shirts and sheets; otherwise they left it alone. They applied no topical antiseptics of any kind.

Rob began to feel a surprising emotion.

"Something of a bond was developing between the fly and me. At least, I felt like I was nurturing a separate life form, and that there was an implicit contract in our relationship: I was not attempting to remove my guest, and the guest was doing what it could to be unobtrusive and to keep the wound infection-free. As I remarked to Nancy, this was as close to pregnancy as I would ever get.

"In the last few days of My Parasitized Life, the larva became very quiet, moved close to the opening of its burrow (where Nancy could see it clearly), and all indications of movement and excretion ceased. Then one day, as I was removing my T-shirt after a sweaty run, I felt it catch on something on my back. I called Nancy. The larva was slowly emerging, hooked ring after hooked ring, and I felt . . . absolutely nothing. Perhaps the larva had anesthetized the surrounding tissue; I can't think of another explanation, because the lack of sensation was complete."

Nancy caught the larva as it dropped, and they kept it in a jar on a moistened paper towel in the dark until the adult fly emerged from its pupal case five weeks later. The fly and its puparium now reside in the AMNH's entomology collection. Rob told me that

The adult male botfly that emerged from Rob Voss's back, perched on its pupal case.

(© DAVID GRIMALDI)

the larva's exit wound healed quickly and is now invisible, whereas Dr. X's incision left an enduring and obvious scar.

As a footnote to Rob's bold venture into botfly husbandry, I contacted Dr. David Grimaldi, an invertebrate zoologist at the AMNH, to examine the specimen and ascertain its sex. He kindly agreed. The botfly was a male and, according to Grimaldi, a robust specimen. "Must have fed well," he concluded.

Death by Decapitation

In an interview for a review of her book *The Secret Life of Flies*, author and dipterist Erica McAlister actually expresses disappointment that she has yet to play host to her own botfly. Now that's commitment. If Dr. McAlister were an ant, however, I expect she would pass on an opportunity to play host to a scuttle fly.

Being among the most opportunistic creatures on Earth, flies have not squandered the chance to exploit one of the most abundant creatures on Earth: the ants. Among the most accomplished of these are the scuttle flies, of which more than 200 species make their living by infiltrating the lives of their distant ant cousins. Some of the most charismatic species in this fly family are those in the genus *Pseudacteon*, known, as we shall see, for their ant-decapitating habit—and for their potential to control invasive fire ants.

I first encountered scuttle flies—so named for their rapid running behavior with frequent abrupt stops—during my monthlong bat-research sojourn in South Africa's Kruger Park in 1985. (Some species are also known as satellite flies, for their habit of hovering at a constant distance and location relative to their prey.)

In the African paradise of Kruger, I had many opportunities to indulge a favorite pastime: watching animal life large and small. By day, we could scan the ruddy waters of the Luvuvhu River for hippos, goliath herons, and the occasional crocodile. At night, large black scorpions and whip scorpions patrolled the driveway as the haunting sounds of distant hyenas and lions wafted across our campsite.

Almost daily, a colony of a few hundred Matabele ants silently wended their way single-file through our campsite adjoining the Luvuvhu. Half an hour later, almost like clockwork, these large, blind social insects returned, following the chemical trail they had left on the outbound journey. This time, however, their jaws were stuffed with the white bodies of termites whose abundant nests they had raided. The ants' orderly convoy evoked the discipline of well-trained soldiers, although I discovered that they would scatter, hissing angrily, if I disturbed them by blowing air. Within seconds they would regroup to resume their silent march.

One day as I was gazing at this funereal procession, I noticed a tiny gray insect perched an inch from the ant column. Closer inspection revealed the bystander to be a fly, and it appeared to be keeping close tabs on the ants. I didn't know it at the time, but it was probably a scuttle fly. Years would pass before I learned that these flies practice an audacious, ant-dependent mode of child-rearing.

Scuttle flies number about 4,000 species worldwide, and many are parasitoids of other insects. The champion parasitoids are the wasps, but the advantages of this survival strategy haven't been lost on flies. At least 28 species of South American leaf-cutter ants alone are known to be targeted by at least 70 species of these flies. A 1995 study found 127 species of ant-parasitizing scuttle flies in just one genus at a field research station in Costa Rica; most of them specialized on a single ant host species.

The flies usually stalk their prey on the trails the ants run along. They dart in on their quarry, laying a single egg, usually on the ant's thorax. Some flies inject the ant with a sharp ovipositor at the rear end. The fly's maggot hatches and, if necessary, penetrates the fissure between the ant's head and thorax. From there the maggot burrows toward the head, which contains the large, nutritious muscles that power the ant's biting mouthparts. There the little invaders feed, carefully avoiding nerve tissue, whose destruction would incapacitate the ant and compromise the maggot's food source. After a couple of weeks, the full-grown maggot releases an enzyme that dissolves the membrane connecting the ant's head to its body, causing the head to drop off. While the headless body stumbles around, the discarded head becomes a protective capsule where the maggot pupates before emerging as an adult fly about two weeks later.

Because an ant's head must be large enough to accommodate a maggot that grows to several times its original size during its residence, scientists have speculated that small size might be advantageous to ants by making them too tiny to be a suitable fly target. But if natural selection is rendering punier ants, the flies may be keeping up. The new title of world's smallest fly goes to a scuttle fly discovered in 2012 in Thailand. Adult *Euryplatea nanaknihali* measures just 0.4 millimeter. "It's so small you can barely see it with the naked eye. It's smaller than a flake of pepper," says Brian Brown from the Natural History Museum of Los Angeles County, who discovered it in a malaise trap (which uses fine meshing to funnel flying insects into a bottle of preservative) in Thailand's Kaeng Krachan National Park during an intensive insect-collecting project. Ant predation by this fly hasn't been seen, but there are some telling clues: its pointy ovipositor is well-suited to the task, and its closest relative decapitates ants in Equatorial Guinea.

Securing a safe haven for one's offspring isn't the only reason a scuttle fly may try to get inside an ant's head. In a 2015 study, a research team led by Brown describes grim but electrifying experiments in which they crush ants with forceps in the wilds of Brazil, then watch and wait. The ants' injuries mimic those inflicted during skirmishes with other ants. Within minutes, one or more tiny flies arrive at the scene, apparently drawn to the odors emanating from the injured ants. Having located her injured quarry, the fly runs in rapid circles around the ant, occasionally darting in to tweak a leg or antenna. Like a curious cat who pokes hesitantly at an unfamiliar object that might be alive, the fly seems to be assessing the approachability of the stricken ant. Misjudgment here could be fatal to the fly, as these ants will easily crush them in their jaws and eat them. If the fly deems the ant sufficiently disabled to preclude effective self-defense, the emboldened fly starts to make brief runs onto and off the victim. Then, if the coast is clear, the fly uses a long proboscis with a bladed cutting tip to saw its way through the ant's neck. Task nearly complete, the fly dismounts, grasps the ant's jaws in its forelegs, and with a few rapid tugs is usually able to detach the ant's head, which is dragged away to a more secluded spot. I cannot watch videos of this behavior without attributing awareness to the minuscule flies.

You may be wondering what these intrepid flies are doing with their ant trophies. The fact that only females engage in this morbid form of predation offers a clue. Of a sample of 16 decapitators, none had mature eggs in their abdomens, and since they could not have already borne young, the research team surmises that the flies eat the contents of the ant head capsules to mature their eggs.

How often do ants' bodies get invaded by scuttle flies? It varies widely by species, location, and time of year. It can exceed one in three ants, but I doubt that higher rates would be sustainable, because flies depend on healthy ant populations to infiltrate. In a 2017 study, scientists collected 89,699 ants of two species from the wild in north-central Brazil, then carefully monitored the ants for harboring fly parasites. Over the next two weeks, several thousand flies began to emerge from their ant hosts. Parasitism rates were 1,042 ants (1.6 percent) of the first species and 1,258 ants (5.4 percent) of the other. In very rare instances—five out of nearly ninety thousand ants—larvae of two different fly species emerged from the same ant. The preferred route of emergence was the most convenient one: the ant's mouth. It isn't known just how the blind larvae know where in the darkness of the host's interior they need to migrate to make their exit; perhaps they use chemical cues. Other exit points have included the ant's abdomen, and the leg. By that time, the ant has died, but only recently; workers of some infected ants were contentedly (as far as one can tell) transporting plant fragments the day before their little Aliens emerged.

Riding Shotgun

Lest it appear that ants are helpless quarry to these flies, let me assure you they are not. Famous for their organization and discipline, some ant societies—including the hyperabundant leafcutter ants—have been evolving orderly defenses against their fly foes.

Ant parasitism by satellite flies is serious enough to have spawned an arms race of sorts. Ants are formidable creatures, and some afflicted species are not taking the fly invasion passively. The appearance of tiny parasitic flies hovering above like diminutive combat helicopters sends the ants into a panic. As the flies seek opportunities to dart in and deposit a single egg on the neck of an ant, the ants raise their heads and gnash their mandibles threateningly. One species has been seen catching flies by flicking its head over its own back and crushing the uninvited guest in its jaws.

During a visit to Mexico in 2008, I watched leaf-cutter ants returning to their nests with their botanical booty and noticed that some of the leaf pieces carried by the large worker ants had tiny worker ants perched on them. I learned later that scientists had been puzzled by this curious and apparently inefficient behavior, until Donald Feener and Karen Moss, at the Smithsonian Tropical Research Institute in Panama, discovered in 1990 that the *minims* (the term given to the tiny hitchhiking workers) were playing the role of guards.

This strategy thwarts a common approach used by the flies: landing on a leaf fragment, then approaching the carrier ant's head to lay eggs, in some cases directly into the preoccupied ant's open mouth. The minims have been known to kill any incautious fly who dares tread where they are posted. That as many as four minims may be seen riding shotgun on a single leaf fragment, each with her menacing jaws wide open, attests to the level of threat posed by parasitic flies. A cynic might suggest that the minims are just being lazy, an idea undermined by the legendary selflessness and epic energy of ants.

In their yearlong study, Feener and Moss found that flies were far less likely to land on a leaf fragment that was guarded by a minim ant, and if a fly did land, it spent less time on the leaf and was much less likely to lay an egg.

Since learning about the leaf-guarding behavior of these ants, I have wondered whether the leaf-cutters' defense is built into their genes or whether it's a more flexible response to the presence of the parasitic flies. Maybe it varies by geographic region, although I suspect that wherever leaf-cutter ants occur there are grateful flies to try to exploit them. If the ants' response to the flies is flexible, then we may expect that the ants would shelve their fly defenses in the absence of flies.

I had an opportunity to shine a small beam of light on this question during a visit to the Montreal Insectarium in July 2018. Among the live displays there is an impressive colony of 15,000 leaf-cutter ants. This display is open and one might easily reach in and touch these industrious insects, which seem either oblivious or habituated to the curious humans watching and breathing nearby. The ants' close proximity allowed me to lean in and inspect leaf fragments being carried from one end of the display to the other. I could detect no minims riding on them.

I also couldn't detect any tiny flies lurking among the ants. Although admittedly they could easily be overlooked, one would not expect to find natural parasites of an ant colony located thousands of miles from its native range. I decided to ask Gabrielle, a young docent who had just finished an interpretive presentation on the leaf-cutters for a gathering of about twenty visitors.

"Do you ever see smaller minims riding on leaves being carried by their larger comrades?"

"I've heard of that as a defense against contaminants, but I don't recall ever seeing it in this colony," she replied.

A sample size of one is hardly a rigorous test, but I am chalking it up as anecdotal support for the hypothesis that defensive behavior against parasitic flies by leaf-cutter ants is a flexible response to the fly threat. That is, the ants carry guards on their leaf fragments only when they need to. When I returned to the Feener and Moss study, I discovered that they had conducted experimental introductions of the fly parasites at ten ant colonies, and found that leaf-cutter ants adjust the level of leaf guarding, sometimes within just 20 minutes, in response to the presence of the flies.

Could it be that the depredations of the parasitic flies have driven the evolution of the tiny minim ant caste? When I posed the question to Cambridge University professor Henry Disney, an authority on the associations between parasitic flies and leaf-cutter ants, he replied that he was not aware of any published speculations about it.

The study of parasitism of leaf-cutter ants by scuttle flies is motivated by more than scientific curiosity. Despite their small individual size, such is the abundance of leaf-cutter ants that they are collectively the most profligate herbivores in the neotropics and are widely defined as crop pests.* Some species in the genus *Atta* can defoliate an entire citrus tree in less than 24 hours. Because the ants cannot simultaneously cope with flies and harvest

*We may bear in mind that leaf-cutter ants were blazing trails millions of years before the invention of chainsaws and bulldozers, and they provide many benefits to healthy rain-forest ecosystems, such as improving soil richness, soil aeration, and sunlight penetration.

leaves for the underground fungus gardens they grow for their food, fly attacks significantly reduce the ants' impact on forests in their range. Research is ongoing on how the flies might be used to facilitate forest regrowth.

Scuttle flies are also being deployed to try to combat the presence of invasive fire ants, which have been marching their way across southern North America, where they have no native fly enemies. In 1997, entomologists introduced ant-decapitating flies from Brazil, and five species have become firmly established as important natural enemies of fire ants in the southeastern United States. However, introducing nonnative species is a risky enterprise. One of the worries with importing ant-targeting flies is that they may begin to target native ants that don't cause ecological damage the way introduced ant species do. There are now documented cases of benign native fire ants being parasitized by introduced flies.

Some scuttle flies have gone beyond attacking ants in the open to dwelling in the nests of army ants, where they scavenge or prey upon their hosts. So successful is this invasion that some of these flies may occur in the thousands in a single army-ant colony. It seems likely that these flies accomplish their feats of infiltration with the aid of chemical secretions that mimic those of the ants. But simply looking like your hosts would appear to be an important part of the ruse, and it is perhaps easier to resemble the relatively amorphous form of a larval ant than to impersonate an adult. The legless, wingless adult females of *Vestigipoda* flies are, to us and presumably to the army ants who feed and tend to them, astonishingly convincing mimics of ant larvae.

While best known for their associations with ants, scuttle flies as a group target diverse prey. They haven't missed the bountiful opportunity presented by other very successful kin of ants, the termites. Female con flies, for example, deceive worker termites into leaving their colony. The fly lands near a termite worker who has emerged from a breached nest, then prods the termite. The prod casts a mysterious spell on the termite, who now follows the fly on a long walk. Far away from the safety of the colony, the termite is somehow immobilized by the fly, who then lays an egg in its abdomen, covers it with soil, then guards her comatose and paralyzed victim. In some termite-specialist scuttle flies, females are wingless, and before entering a termite colony they are carried around by winged males while in copulation.

Fly Catches Hummingbird

I admire the temerity of small flies that attack ants, but if there is a Rolls-Royce of flies, my vote goes to the family Asilidae, commonly known as the robber flies. These big, burly, large-eyed, and usually hairy aerial predators are the dipteran answer to avian flycatchers. The largest of robber flies have on rare occasions caught and overwhelmed small hummingbirds. Like a mother botfly seeking a suitable mosquito courier, robber flies use an ambush hunting strategy, perching on the ground or low-lying vegetation then zipping out like a guided missile to nab any suitable-looking prey that happens by. They do not attack from behind, but instead intercept the prey by anticipating its location, as a quarterback does a receiver. But there will be no touchdown celebration for the hapless target, who will most likely be dead within seconds (see the image in the photo insert).

Robber flies' piercing bites inject venom, which quickly immobilizes the victim, allowing the fly to use its jaws as a straw, sucking out the liquefied innards. Because insect blood does not flow through vessels like ours does, any portal into the prey will allow the fly to feast. Most robber flies have a beard of stiff hairs just below the eyes, which appear to help protect the eyes from damage during a captured prey's brief few seconds of struggles before the paralyzing saliva takes effect.

Though they are not especially common, I have seen many robber flies in nature. They were seasonally fairly abundant at a small, protected dry-scrub habitat that I often visited while living in south Florida (Seacrest Scrub Natural Area, in Boynton Beach). As I walked the trails on any given day, I might see a dozen or more robber flies, mostly perched on the sandy path. They are wary and hard to spot until they fly, but they usually land again a few yards ahead, and a patient observer with a chameleon's approach can get close enough to photograph them.

On one occasion I happened to notice a robber fly perched on a leaf about two feet off the ground. With my binoculars I was able to watch the fly from outside her flee-zone. I waited for seven minutes. The fly remained completely motionless. Theirs is a patient art, and I, like most humans, had things to do. I was outwaited on this occasion. I've discovered robber flies consuming prey, but I've yet to see a live pursuit and capture.

Robber flies, of course, have enemies of their own, which may explain a strange behavior in some species: dangling from vegetation by one foot while dining on, say, a bee or dragonfly. It isn't known why they do this, but I wonder if it is a defense against ants, whose constant wanderings are less likely to intercept one leg than several.

Despite their formidable appearance—the largest species, Madagascar's giant robber fly (*Microstylum magnum*), measures 5.7 centimeters (2¼ inches), and another large contender bears the name *Satanas gigas* ("Great Satan")—robber flies are not aggressive toward us. Although they are renowned for tackling prey larger than themselves, we are way too big to qualify for their menu. During a conversation in a sports bar with a couple of professional entomologists who have studied these flies, I asked if they had ever been bitten. Tristan McKnight, who studies robber flies in the Department of Entomology at the University of Arizona, chimed in with a characteristic entomologist's perspective:

"The only way I can get them to bite me is to hold them between my fingers and press them against my skin, and even that doesn't always work." (Chalk up a nobility point to the robber flies there.)

After a tug on his beer, he added, "It's not very painful, nothing like a wasp sting, probably because it's not a defensive bite. It's just a bite meant to subdue prey and inject them with saliva that digests rather than stings."

McKnight is not alone among scientists for being curious about the virulence of insect bites and stings. When I followed up with him via email, he told me he has been forcing various robber flies to bite him as part of a fly extension of a four-point pain scale for wasps and bees painstakingly developed by the acknowledged "King of Sting," Justin O. Schmidt, another curious entomologist based in Arizona.

McKnight directed me to an Instagram post describing his experience with one of Michigan's largest robber flies, a 3.3-centimeter-long (1¼-inch) *Proctacanthus*.

"They're definitely not wimpy," says McKnight. "My verdict: a solid 2 [Schmidt's equivalent of a honeybee sting]. The bites hurt like a needle stab, and then quickly swelled up into 4 mm wheals with a 10 mm flare and an itchy burn. The throb of pain and raised wheals [easily visible in McKnight's photos] had mostly subsided within 35 min, but the . . . arm stayed hot, red, and swollen all night and into the next day." It is, I suspect, thanks to the likes of Tristan McKnight that there is actually a World Robber Fly Day, the last day of April.

Fighting Back

As you can imagine, many robber-fly victims are other flies. And like most targets, attacked flies evolve defenses. I don't know of any studies of defense against robber flies, but the defenses of fruit flies against wasps are another story. Fruit fly larvae are commonly infected by parasitoid wasps, who inject eggs into the seemingly helpless grubs with a needle-like ovipositor. But the maggots are not oblivious to their assailants. At an entomology conference in Vancouver, I watched close-up videos of the diminutive, shiny black wasps poking their syringes at the creamy maggots through a thin layer of mushy fruit. In response, the maggots performed two types of evasive behaviors: curling or rolling. To curl, the maggot quickly flexes its body into a C shape. To roll, the maggot twists its body along the long axis, lurching to the right or left. Both behaviors generate a moving target for the wasp, who often gives up and moves on to seek easier prey.

Even if the wasp succeeds in harpooning her quarry, it isn't a checkmate for the wasp. Infected flies mount an immune response termed *cellular encapsulation*, in which fly immune cells form a multilayer capsule that entombs the implanted egg inside a hardened, impermeable shell. So estranged is the egg from the rest of the fly's bodily functions that the dark, dormant oval mass it becomes is still visible in the adult fly's abdomen after its innards have undergone the dramatic rearrangement during larva-to-adult metamorphosis. In its cocoonlike tomb, the developing wasp presumably dies due to toxicity, asphyxiation, or physical entrapment.

This is another arms race of sorts, with ways of living at stake, and the parasitoids aren't staring blankly at the flies' ramparts. In small organisms that evolve rapidly due to short generation spans, adaptive traits may arise more quickly in a dynamic process of push and push back. Wasps are evolving chemical virulence factors that suppress the flies' cellular encapsulation.

Fruit flies have yet another way to resist parasitoid wasps. Probably due to their habit of feeding on fermenting fruit, they have developed a tolerance to alcohol, and it turns out to be a useful tool against enemies. Fly larvae who have imbibed are less likely to be assailed by wasps. Alcohol consumption can even zap developing wasps in already infected fly larvae. And the diminutive flies appear to know it. "Infected fly larvae actively seek out ethanol-containing food, showing they use alcohol as an anti-wasp medicine," notes Todd Schlenke of Emory University, who has studied the phenomenon. "These little fruit flies, the same that hover around the brown bananas in your fruit bowl, are making complex decisions about how much alcohol to consume based on whether or not they have internal parasites."

Adult fruit flies are in on the game, too. Parasitized females of several species give their offspring a leg up by preferentially laying eggs on food containing alcohol. This "prophylactic kin medication" is not practiced by uninfected females. Female flies recognize their wasp parasites by sight (not smell), and they gravitate to alcohol only when exposed to female wasps (up to four days earlier) but not benign male wasps.

The phenomenon meets the four scientific criteria for self-medication:

1. The substance in question, alcohol in this case, must be deliberately contacted.
2. The substance must be detrimental to one or more parasites.
3. The detrimental effect on parasites must lead to increased host fitness.

4. The substance must have a detrimental effect on the host in the absence of parasites.

That last criterion adds stringency by assuring that the flies' preference for alcohol is not merely a dietary choice. The only other insects currently known to self-medicate are honeybees and a handful of butterflies and moths. The phenomenon is more widespread in vertebrate animals, enough so for it to have been given its own name: *zoopharmacognosy*—from the Greek words *zoon*, "animal," *pharmakon*, "drug," and *gnosis*, "knowledge." Still, there are some who will question whether a fly knows what it is doing and why.

Parasitoid defense has led to at least one other interesting discovery in fruit flies. Researchers at the Geisel School of Medicine at Dartmouth College discovered that adult flies, when threatened by wasps, send warning messages to other flies with rapid wing movements. The fruit flies fear the wasps so much that when they spot one, they cut future losses by laying fewer eggs. Even flies who have never encountered a wasp before lay fewer eggs on hearing their comrades' warning whines. Closely related *Drosophila* species produce slightly different warning sounds, but just as birds learn the alarm calls of other species they forage with, so do the flies learn to recognize the warning dialects of their neighbor species. "The dialect barrier can be alleviated through socialization between species, without which, information would otherwise be lost in translation," says study author Balint Kacsoh.

If you can't avoid predatory wasps in your midst, why not just look like them? Thousands of fly species have evolved as doppelgängers for wasps and bees. In addition to the possibility of being overlooked by their stinging foes, these mimics repel larger predators—a blackbird, say, or a flycatcher—that wish to avoid being stung. Many if not most of these mimics don't settle for looking like any old bee or wasp. There are flies that mimic yellowjackets, bald-faced hornets, potter wasps, and ichneumon wasps (see the pair of images in the photo insert). Suitable behaviors strengthen the illusion. While photographing flies in Ecuador, Steve Marshall encountered an unnamed hoverfly species exploring a honeydew-spattered leaf with a group of almost identical stingless bees. Marshall noted that the fly enhanced the impersonation by doing something beelike but unflylike: dangling its hindlegs in flight.

You might have noticed a category of parasitic flies conspicuously absent from this chapter. I'm referring to flies with a specific culinary target: blood. So important is the impact of blood-seeking flies on the humans and other creatures they hunt that they merit a chapter of their own.

Chapter 5

Blood-Seekers

How fares it sadly with the man
Whose soul doth patience lack
When he to smite fugacious flies
Himself doth fiercely whack.

—ANONYMOUS

Flies are among the most intrusive of all insect groups thanks to their penchant for sharing our homes and invading our bodies (not to mention transmitting devastating diseases, which we will explore in chapter 10). But for all their annoying accomplishments, it is their particular habit of stealing our blood that trumps all others. In addition to rhyming verse, flies' lust for blood has fueled our own lust to repel, evict, and terminate them.

The human-blood-seeking lifestyle places those flies that practice it in a minority ecological category: those who benefit from us. It is widely acknowledged that the modern human presence on Earth is detrimental to most other species. Urban encroachment, climate change, biodiversity loss, a looming Sixth Great Extinction—these are burdens of the Anthropocene Epoch. But with a total skin surface of nearly 12,000 square kilometers

(4,600 square miles),* humans provide one of the world's largest single-species feeding habitats for blood-sucking creatures. Only cows and goats rival us.

I've wondered how many irritated or casual waves, swats, smacks, blows, and splats must occur each day at the human-fly nexus. Surely mosquitoes and other biting flies have plagued humans for as long as humans—including our now-extinct relatives *Homo erectus*, *H. habilis*, the australopithecines, etc.—have walked the Earth. During their explorations of South America from 1799 to 1804, Alexander von Humboldt and his French co-explorer Aimé Bonpland endured constant bites and swollen, itchy skin. They coughed and sneezed, as the simple act of breathing drew the insects into their mouths and noses. Bonpland, who was in charge of the botany collection, took to using the natives' *hornitos*—small windowless chambers they used as ovens—as a means of escape, the high temperatures and smoke still being preferable to the biting blight outside.

You don't have to be an explorer to know the vexing ways of biting flies. Most humans have encountered the mosquito's cunning, and many of us also know the blackfly, the biting midge, the deerfly, and the horsefly. How do they succeed at such an audacious and seemingly perilous lifestyle? Let's find out.

---

*Average human skin surface area is about 1.75 square meters.

Multiply by 7 billion = 12.25 billion square meters.

There are 1 million square meters in a square kilometer.

So there are 12.25 billion / 1 million = over 12,000 square kilometers of human skin on Earth.

There are 2.59 square kilometers in a square mile.

So there are 12 million / 2.59 = over 4,600 [4,630] square miles of human skin on Earth.

Winged Victory

The name Winged Victory applied to a fly comes from a 1923 poem by D. H. Lawrence simply titled "Mosquito." It's a fitting label. There are about 3,568 described species of mosquito. They are found on all continents save Antarctica, and they number some 110 trillion individuals on Earth at any one time. That's about 15,000 mosquitoes for every one of us. On my visit to Montreal's Insectarium, I learned that there are 57 mosquito species in Quebec alone! Luckily, the great majority of them don't seek human blood, and of those that do, it's only the females. Given the mosquito's omnipresence, these are small consolations.

The word *mosquito* derives from the Spanish for "little fly." In addressing his mosquito, Lawrence guessed that her long legs help her land more softly, "And weigh no more than air as you alight upon me, / Stand upon me weightless. . . ." While many flies are smaller, the slender, wispy build of mosquitoes may be an adaptation for landing on their prey with minimal force. Being light has the further advantage of reducing ballast after you have filled your tank with fuel.

Collectively, mosquitoes are very successful at attaining that goal. Some 1.6 million gallons of American blood is sucked by mosquitoes each year, according to an article promoting Zevo, an indoor plug-in insect trap from Procter & Gamble, though I caution that advertisers are notorious for exaggeration.* Someone who may equally be called brave or foolish once ascertained that in the interior of Alaska, a 56-square-inch area of human forearm

*Before you conclude that these zappers are effective mosquito-control devices, you should know that studies have repeatedly found them to be not only ineffective but possibly even countereffective.

can receive 280 mosquito bites per hour. You may be relieved to know that that's an estimate, so they must have extrapolated from a briefer sampling session. You may be further relieved that it would take somewhere between 200,000 and 2 million mosquito bites to kill you from blood loss. That works out to a few hundred bites for every square inch of skin.

How do they do it? The business end of a female biting mosquito is a marvel of evolutionary engineering. Hidden within the visible proboscis are sets of blades, needles, and tubes, all of which work in coordination to achieve penetration of your skin, laceration of a small blood vessel, injection of saliva containing a blood-thinning agent, suction of blood, and, task complete, withdrawal of all this equipment prior to takeoff. For members of the *Anopheles* genus, skin penetration is achieved by a pair of reciprocating saw blades that penetrate by sliding against each other like a pair of electric carving knives.

However much hatred one may muster for their fiendishly irritating ways, we might reserve some admiration for an animal whose task is to approach, harpoon, and remove blood from a large, sentient, alert mammal. It doesn't help them that their flight speed is typically only about 3 miles per hour. Mosquitoes have the additional liability of producing what to us (and we can only assume to other host species) is an extremely annoying whine. (Hint: More on this when we get to the mating habits of flies in chapter 8.) It will be a great day for mosquitoes when they invent the silencer.

For all the hazards it presents, the mosquito's bloodsucking lifestyle is clearly very successful, to judge from the abundance of mosquitoes on our planet. Their success may be more attributable to the prodigious number of babies they produce than to the adult's persistence. According to one study, there may be over 1,390 *Culex* mosquito larvae per square foot of water in a Korean rice field. I couldn't help noticing that this leads to the alarming possibility that certain rice paddies could theoretically produce more biting insects than grains of rice. We may give thanks to the depredations of mosquitofish, betta fish, and other mosquito-feeders of rice paddies and other mosquito haunts.

Mosquitoes' abundance also owes itself to their habit of laying eggs in temporary aquatic habitats. Previously confined to living in ephemeral puddles and relatively rare tree holes, mosquitoes' aquatic larvae have adapted easily to life in tires, rain barrels, flowerpots, beer cans, and other water-holding containers, even baptismal fonts. Plastic garbage has also greatly expanded breeding niches for these insects.

These niches are nearly devoid of predators save for the occasional predatory beetles that may pop in and out. Mosquitoes also have drought-resistant eggs, which they prefer to lay on soil that still bears the smell of larvae from previous generations. That smell can last for years, allowing viable eggs to accumulate during dry years, then hatch all at once in a wet year.

Far more troublesome than their ability to home in on larvae is their ability to home in on us. They use a range of cues to do this. They are sensitive to our movements, our body heat, our odor, the water vapor from our sweat, and the carbon dioxide from our breath. (I have experimented with holding my breath while hiking in the woods. Results are so far inconclusive and effectiveness temporary at best.) It is not unheard of for a mosquito to feed on the still-warm blood of a recently deceased animal. Mosquitoes even have special sensory structures that can detect a host's emissions in the infrared light spectrum, an ability exploited by infrared-light insect traps. Emitted infrared light reflecting off water helps mosquitoes to recognize potential breeding areas.

However they find us, some of us seem more findable than others. Chances are you've heard a fellow human claim—with a blend of pride and peevishness—that they are especially attractive to biting flies. Perhaps you are one of these perceived mosquito magnets. According to a 1966 study, mosquitoes appear to be more attracted to men. There is also evidence that mosquitoes are attracted to dark clothing, movement, sweat, and beer drinkers. So next time you are in mosquito country, wear something white and encourage your buddies to do some exercise to work off their next beer.

Somehow, the idea has propagated that a mosquito can get drunk by feeding on an alcohol-impaired human. There is little evidence to back it up, probably because the actual amounts of alcohol ending up in the mosquito are minuscule. The little biters also shunt impurities into a separate digestive pouch, where enzymes break them down.

Whether male or female, drinker or abstainer, a mosquito magnet might be well advised to swat with determination. A recent study of the mosquito *Aedes aegypti*—a human specialist—found that it can remember the turbulent experience of a violent swat that missed along with the associated scent, and it will move on to safer quarry. Researchers put the insects through tests involving tiny flight simulators, mini wind tunnels, and the odor of humans, rats, and chickens.

What if their preferred prey, attractive or not, just isn't available? Research has shown that mosquitoes that specialize on humans will, in a pinch, shift their attention to, say, dogs or cattle—but given the chance, they will gravitate back to us. There's good reason for host choosiness by mosquitoes, because what kind of host you feed on matters, at least for some parasitic flies. While the nutritional needs of bird-preferring *Culex quinquefasciatus* mosquitoes are met by feeding on either bird or mammal blood, captive mosquitoes fed exclusively on bird blood have higher fecundity (eggs laid) and fertility (reproductive success) than mosquitoes relegated to mammal blood.

Mosquito larvae use the same feeding method as do the largest animals on Earth, the baleen whales. Both are filter feeders, straining food material and regurgitating the filtered water. The keen-eyed wrigglers—who, also rather like whales, breathe at the water surface through a snorkel on the head—are as alert to danger as are the adults. As many a curious naturalist will have observed, they will scamper to the bottom at the sight of overhead movement or moving shadows.

Only female mosquitoes have the mouthparts for sucking blood, which they use more as a source of protein for their eggs than for personal nourishment. Males seek nectar and other plant sugars, with which some females may also supplement their diets. Male mosquitoes of a few species do approach humans or other warm-blooded animals, but not to feed on them; they are awaiting the arrival of hungry females to mate with. Otherwise males of most species swarm.

More scientific interest surrounds the fluids mosquitoes inject into our bodies than the red stuff they take out. Mosquito saliva contains anticoagulants, which facilitate siphoning; bloodsucking would be impossible without blood thinners, which prevent the proboscis from clogging with clotted blood. You may not have noticed it, but most of us become less reactive to mosquito saliva as the biting season progresses. This is an immunological response: more exposed, less reaction.

Fly guts contain many capacious diverticulae, branching pouches that allow the insect to ingest and store large quantities at one sitting. Your uninvited guest may extrude one or more drops of reddish urine while she feeds. This is mostly water, and it allows the insect to acquire a more concentrated blood meal. Even though a mosquito sucks up two to three times her weight in rich blood, it still amounts to only some five thousandths of a milliliter (one thousandth of a teaspoon). If the mosquito gets away with it, take heart that you might have just enriched a lizard's or a spider's next meal.*

The itch felt after a mosquito bite is due to salivary proteins left at the crime scene. If saliva were the only thing mosquitoes left us with, they wouldn't be such a problem. Mosquito biting mouthparts are too small and clean, and too rarely used to transmit illnesses or diseases like cold viruses or AIDS/HIV through simple contamination. The serious mosquito-borne diseases are those that have evolved special ways to avoid the mosquito digestive tract and colonize the salivary glands, from where they can be transmitted through the bite. It is not so much for their bite per se, but for the deadly diseases they transmit that mosquitoes are so closely studied by us. There are entire journals on the study of mosquitoes. At the Mann Library on Cornell University's campus I found: *Mosquito Systematics*, *Mosquito News*, the *Journal of the American Mosquito Control Association*, and a bound collection of reprints spanning 1896 to 1956.

When all is said and done about mosquitoes, what I find most enigmatic about them is that they can—despite what seem to be overwhelming odds against them—outwit us. Or simply outlast us. Tormented in the middle of the night by a lone mosquito in my room, I have on occasion resorted to a technique of surrender. Unable to nab my assailant, and unwilling to rouse myself into complete wakefulness by switching on a light to launch a murderous pursuit, I have waited for the mosquito to alight on my skin

*Mosquitoes have many enemies, including water striders, aquatic beetles, dragonflies, ants, birds, bats, lizards, and frogs. They also have their own parasites, such as nematodes ingested by the hungry larvae. These tiny parasites soon grow to four times the length of their hosts, whose body they consume from the inside. They emerge leaving behind an empty husk.

and take her fill. Sated, she loses interest in me, and I can sleep in peace. Sort of.

Little Biters

For all the triumphs of mosquitoes, there are no lesser marvels in the dipteran realms of blood-seekers, and some are much smaller. The most diverse of these are the biting midges, more familiarly known in North America by their various nicknames, including sand flies, no-see-ums, and punkies. There are over 6,200 known species of biting midges in the world. The family name, Ceratopogonidae, derives from the Greek *keratos*, "horn," and *pogon*, "beard," referring to the males' erectile antennal plumes, which, when the males are not courting females, are flattened against the antennae, giving them the appearance of hairy horns.

But it's not the horny males that get our attention; it's the females lusting for our blood. In a study of neotropical no-see-ums, 70 of 266 *Culicoides* species had been known to feed on humans. Fortunately, only a handful of these—bearing such names as *C. phlebotomus*, *C. insinuatus*, and *C. pseudodiabolicus*—are significant pests.

Biting midges have the most diverse feeding habits of all biting insects. Of those that stalk vertebrates, their menu is broad, with many species of mammals, birds, reptiles, and amphibians, and at least one fish—the air-breathing mudskipper of Malaysia—on the menu. Others seek smaller fare, plugging into the wing veins of dragonflies, damselflies, lacewings (see the image in the photo insert), moths, or butterflies, or other parts of katydids, stick insects, crane flies, spiders, stinkbugs, or caterpillars. Like ticks, they become so engorged that their abdomens may swell to many times their normal size, their legs, head, and thorax resembling appendages stuck to a hairy melon.

During a Skype interview, Art Borkent—with whom we'll get better acquainted shortly—held up a photo of a dragonfly, its wings flecked with about 170 biting midges. Borkent thinks that these tiny freeloaders get more than a meal from their large insect host.

"I think the dragonflies may act as useful dispersal agents for the midges," Borkent told me. "Some dragonflies travel up to 100 kilometers [60 miles] from their natal ponds. That's cheap airfare for little flies that also breed in aquatic habitats."

At least their targets get to live another day. Other biting midges that prey on insects their own size use more lethal methods. A female will fly into a male swarm of nonbiting midges, grab one, inject him with a dissolving enzyme, then suck out his insides. Others go for more direct cannibalism, consuming their mate during copulation. While coupling, the femme fatale impales the male, usually through the head, and injects him with enzymes to liquefy his contents before sucking him dry. When she drops the dried husk of her former lover, his genitals—still firmly clasping hers—break off, forming an effective barrier against future suitors. Talk about commitment!

Most biting flies take 2 to 5 minutes to draw their blood meal before escaping. Some are slower, at 5 to 12 minutes. At this rate, risk of detection and looming death would seem greater. But the minuscule size of many species—they're not called no-see-ums for nothing—helps them avoid detection. Many can get through hair and fur. They are also aided by their subtle, anesthetic touch, as vouched for by the Canadian novelist Margaret Atwood, who spent formative stretches of her early years in the wilderness of northern Quebec and northern Ontario, where her entomologist father was posted:

"In the woods, you wore pants not because it was butch but because if you didn't wear pants and tuck [them] into your socks you would get blackflies up your legs. They make little holes in you, into which they inject an anticoagulant. You don't feel them when they are doing it, and then you take your clothes off and find out you are covered in blood."

Unfortunately, the numbing effect is temporary. In time, the bites of blackflies and other biting midges can leave disproportionately large and itchy welts. On the bright side for northerners, temperate biting midges do not transmit diseases, unlike their tropical cousins. Nor are blackflies bound to bloodthirsty habits; both sexes also feed on nectar, pollen, and the honeydew excreted by aphids and other insects. But you wouldn't know it during Canada's blackfly season.

Like their mosquito kin, blackflies live in water before emerging as flying adults. Larvae living in flowing streams use hooks to cling to pads of silk that they plaster onto rocks. Predators lurk here, and some types of blackfly larvae have a nifty escape option: they tether themselves to anchored silken threads, allowing them to let go and drift temporarily downstream to escape predators. Once the coast is clear, they clamber back up their silken rope.

The pupal stage, spent glued to submerged rocks, has no such escape mechanism, and many fall prey to predatory dance-fly larvae. Some dance fly grubs will even use the now empty blackfly pupal cases as their own pupation chambers. Emerging blackfly adults that evade detection float Rapture-like to the surface in a bubble of air that explodes to release a winged adult as it pops above the water surface.

Frog Biters

One particular group of biting midges has adopted a curious target: frogs (see the image in the photo insert). When I associate frogs with flies, it is usually the fly that is the frog's meal, but with the frog-biting midges (family Corethrellidae) the tables are turned, albeit not lethally.

If you wish to learn about frog-biting midges, Art Borkent is your man. I met Art in Vancouver at the joint meeting of the Entomological Society of Canada and the Entomological Society of America. Borkent's passion for flies runs deep. He was catching and rearing midge larvae by age thirteen, and he says this interest contributed to some false starts with potential girlfriends.

Borkent—whose formal affiliations are research associate of the Royal British Columbia Museum and the American Museum of Natural History but who works independently for virtually no pay—has an infectious enthusiasm for his chosen field, and when he talks flies he is lively and animated. He and his wife, Annette, who accompanies (and funds) him on most of his field trips, are intrepid explorers. In 1993, they packed their three children in their old Volvo station wagon, hitched a small trailer to the back, and drove from British Columbia to Costa Rica. While the kids went to a local school, Art spent nine months conducting field work on biting midges, frog-biting midges in particular. Annette has provided technical support for practically all of Art's papers; in one of them, a treatise reporting on a seven-week expedition to Western Australia in 2001 to find a rare biting midge, she is pictured close-up as several of these tiny flies settle in for a meal on her eyelids and cheek.

There are currently 113 described species of frog-biting midges (up from 97 when I began researching this book) and, based on limited sampling effort and novelty rates, at least another 500 as yet undiscovered species. They are found mainly throughout the world's tropics; just one species ranges as far north as Canada, and their southern limits include Buenos Aires and New Zealand.

With just a few dozen specimens scattered in collections

throughout the world, little was known of these midges until the mid-1970s, when an American entomologist named Sturgis McKeever got an idea while he was doing field research in the state of Georgia.

It had long been known that the aptly named frog-eating bat eavesdrops on frog calls to find its prey. On a hunch that midges might be doing the same thing, McKeever broadcast prerecorded frog calls in the field. "He collected bucketloads of the midges," Borkent told me. By playing recorded frog calls over a trap—a fan blowing into a net—many specimens could be collected in a single evening. On one of their first forays, McKeever and a colleague caught 566 midges in 30 minutes. Suddenly, biologists had a tool—a cassette tape recorder—to collect and document a group of flies that for a century had languished in obscurity. Apparently, the elusive males, which don't hunt blood, are either deaf or uninterested in frog calls, for the frog-call traps have only ever caught females.

Once a suitable frog is detected by its caller ID, the midge begins to sneak up on her quarry. She flies in short bursts during each call, landing when the frog goes silent, then jumping in the air to resume flight during the next call. Her relatively thick midfemur facilitates a good jump. Once within a foot or so of a frog, the midge switches from detecting calls to detecting carbon dioxide from the frog's breath. If the frog is on dry land, the midge walks toward the frog; if the frog is floating, the midge assumes a bobbing flight, daintily if haphazardly striking the surface until landing on her host. Once she has taken her fill, the swollen midge waddles off the frog, too heavy to fly. She must first cast out enough water droplets filtered from the blood in her belly before she is light enough to take wing.

It is not clear by what mechanism midges hear frog calls. A good possibility is that the calls are detected using the Johnston's organ, a cluster of neurons at the base of the midge's feathery antennae that respond to deflections of individual hairs. It's thought that these organs detect subtle, sound-based air vibrations. The antennae are incredibly sensitive, responding to deflections of less than a thousandth of a degree, about sixty thousand deflections per centimeter.

The flies' bites are not generally dangerous to the frogs, unless they are deluged by flies. One estimate finds that a small frog could lose nearly a third of its total blood volume to a high volume of these flies in an hour. Some, and possibly all, frog-biting flies also transmit trypanosomes—single-celled parasitic protozoans—between calling frogs, and this, like frog-biting itself, is believed to be an ancient association. While closely related trypanosomes cause us sleeping sickness in Africa and Chagas disease in South America, these parasites aren't known to be lethal to the frogs.

Nevertheless, the frogs have not been passively submitting to the abuse. When attacked, they may flick, rub, or swipe flies from their bodies. The frogs also produce a wide range of chemicals on their skin, some of which almost certainly evolved to deter biting flies, of which the midges are the most prevalent. Borkent thinks that by chorusing in groups the frogs might also be confusing the flies. Some male frogs refrain from calling altogether, stationing themselves near the vocalizers and hoping to intercept an approaching female. Other frogs call above 4,000 hertz, beyond the hearing sensitivity of the midges. Yet another anti-midge tactic has been to move far from where the midges breed, including higher elevations. Simply calling while partially submerged also lessens a frog's vulnerable surface area.

A more recent adaptation may be to move into cities, where male frogs have developed more complex calls that are more attractive to females. Producing such calls presents a dilemma for the frogs, for they are also more attractive to the flies, but in urban settings, biting insects and predatory bats are scarcer.

When scientists transplanted the city-slicker frogs into rural areas, they lowered their vulnerability to enemies by simplifying their calls, whereas country frogs were unable to jazz up their calls after being transplanted to the city.

The frogs are not the only ones producing courtship songs; frog-biting midges use "songs" of their own, generated by rapid wingbeats, to attract mates. This raises a question about origins: did the flies evolve courtship songs first, then expand their acoustic fluency into frog detection, or did they leapfrog from eavesdropping on amphibians to mate attraction? Co-opting an organ to perform a secondary function is a widespread phenomenon in nature; our tongues evolved first to allow us to taste things before adopting the arguably greater function of speech. For many mammals, a tail doubles as a signaling device and a fly swatter. Various fishes use their swim bladder—evolved to assist buoyancy—as an organ of sound production. Flies' use of sound in mating is much more widespread across several families of mosquitoes and biting midges than its use in frog eavesdropping. My hunch is that midges listened for mates before they listened for frogs, and that acoustic mate-finding skills were later recruited for frog finding.

Fly Bites Dinosaur

Blood-stealing is not a new thing for flies, and various lines of evidence indicate that frog-biting midges have been feeding on calling frogs since at least the early Cretaceous Era. The oldest fossil records of frogs are from the early Jurassic Period, about 200 million years ago, and the earliest fossil of a frog-biting midge dates to a 127-million-year-old hunk of Lebanese amber. Evidence from a rich fossil record of the frog-biting midges' sister groups—the phantom midges and mosquitoes—and from both fly and frog anatomy suggests that these midges and their hosts have been interacting for at least 190 million years.

Might some of these biters have been stalking bigger prey back then? What about dinosaurs? Borkent described for me a piece of sleuthing he did with biting midges preserved in amber with a *Jurassic Park* twist.

In studies dating back to the 1970s, Antony Downes, an entomologist with Agriculture Canada, began to notice a consistent pattern: midge species that sought the blood of larger animals had fine-toothed mandibles (cutting appendages of the mouth) and fewer sensory hairs on the maxillary palps (a pair of antennalike sensory appendages associated with the mouth) than those that targeted smaller quarry. These hairs function as carbon dioxide detectors.

"Presumably, you don't need to grow lots of these hairs to detect the odor of a huge beast," Borkent told me, "whereas picking up the scent of a little mouse or bird is harder. Blood-engorged midges preserved in 78-million-year-old Canadian amber are not rare," Borkent continued, "and some of the ones I've examined have few of these hairs—a strong indicator that they went after giants. And since there were no large mammals plodding about while the dinosaurs were at the helm, it's a good bet that these midges were sniping duck-billed dinosaurs and tyrannosaurs."

Is any of this preserved blood viable? The core question here is whether DNA was preserved, a hot topic during the time *Jurassic Park* appeared as a movie. Michael Crichton's 1990 book, on which the film was based, relied on the premise that DNA could remain viable for eons in the stomach contents of a bloodsucking fly preserved in amber. Disappointingly, it has since been shown that even under ideal conditions DNA completely degrades in one to two million years.

Maybe there will turn out to be another way to resurrect

dinosaurs in flesh and blood, but it doesn't look like biting flies are going to furnish the goods.

Bitten for a Cause

Mosquitoes, blackflies, and biting midges may be the most notorious of the blood-seeking flies, but they haven't cornered the market. Just as there are smaller biters, there are also bigger and faster ones. And there are bloodsucking flies that have become so deeply enmeshed in the parasitic lifestyle that they have ceased to resemble flies.

As a small personal sacrifice for the research of this book, I decided to let a horsefly bite me. I had felt the initial bite of many a horsefly, but once discovered, she was never allowed to complete her meal. These brief encounters taught me that the pain of a horsefly bite is proportional to the fly's size. Therefore I imagine that it's quite rare for a horsefly to successfully feed from a human, equipped as we are with sensitive skin and slapping and swatting hands that can reach anywhere on our body surface. No wonder they're skittish, often taking off just after they have landed, and generally quick to depart when the affronted's hand approaches.

I got my opportunity on a hot July day northwest of Orillia, Ontario. My winged assailant was rather small as horseflies go, but still about twice the size of the deerfly circling my head. I detected the horsefly as she began to work on my right calf. As her mouthparts drilled a millimeter or so into my flesh, I felt pain but less than I had anticipated. Having fed for about two minutes, she withdrew her mouthparts, shifted a centimeter, then began drilling again. A small spot of blood appeared at the first wound. I watched and waited while she drank from the second location, then a third. Her abdomen pulsed rapidly and her eyes, banded with iridescent green, shimmered in the direct sunlight. She ejected a tiny orb of fluid from her rear, as mosquitoes do to help concentrate the blood. As a droplet of blood from the second bite wound began to trickle down my leg and she brazenly prepared to make a fourth entry hole, I called it quits and shooed her away.

All in all, it was an underwhelming experience. On a scale of 1 to 10, I would rate the pain about 4. With no itchy welt in its wake, I rank this bite below a mosquito's for overall unpleasantness.

A week later, during a break from snorkeling on the shores of Lake Ontario, I took advantage of an opportunity to receive the bite of another common winged blood-seeker: the stable fly (see the pair of images in the photo insert). Anyone wishing to hone their reflexes should spend time with these crafty little biters, and I recommend livestock barns, for which they are named, or beaches in midsummer as good bets to encounter them. Superficially resembling their close relatives the houseflies, both males and females feed on blood, using a stiff bayonetlike proboscis with a sharp, tooth-lined tip. They are devilishly fast; they make the housefly seem sluggish. Their wicked speed and difficulty to swat has been attributed (in part) to the fact that they do not bite by sinking their mouthparts deep into their prey in the manner of mosquitoes. Instead, their saw-tipped bayonet remains near the surface, allowing them to escape in an instant. If you encounter a stable fly, you'll probably have plenty of opportunities to appreciate this, because their bite rarely fails to cause sharp pain, and one's most vehement swat rarely works to discourage another attack.

The pain from this particular fortunate fly who got to feed undisturbed was, like the horsefly bite, anticlimactic. Perhaps there is a psychology at work here, in which pain is less felt when we steel ourselves for it, whereas the sudden discovery of a fly sinking its mouthparts into our flesh amplifies the ouch. However, human studies of heat pain have found that pain sensations are heightened when we are expecting them. I'm hoping some probing dipterist will put my opposing hypothesis to the test someday.

If the superquick stable fly has an opposite, I would nominate a species I encountered on a five-mile hike in 2018 around one of the Green River Lakes in Wyoming's Bridger National Wilderness. My companion and I were visited by many smallish gray-and-black flies with a taste for blood. I later identified them as Rocky Mountain bite flies from the snipe-fly family Rhagionidae. What most struck me about these bloodthirsty companions was their almost complete disregard for my defenses. Now, such is my reverence for life that I shy away from killing insects unless I am under persistent attack. But such was the steadfast persistence of the snipe flies that after a while I began to dispatch a few. I was surprised by how easy this was accomplished when I casually squashed one between my thumb and index finger without stealth or speed. Intrigued, I found I was able to gingerly pick them up, legs wriggling, with the same approach. I could also gently pin them with a fingertip as they trotted across my skin or shirt seeking a good spot to feed. I let one work her jaws into my flesh for a few seconds, and found that she couldn't be coaxed loose when I pushed against her with my finger; she had to be forcibly evicted. Wondering whether their laissez-faire approach to life and death might be a species-wide trait, or whether the flies at this particular location and date were somehow stupefied, I asked my ever-helpful biting-fly expert, Art Borkent.

"Rhagionids are stupid," he said. "You don't have to swat them; you could simply apply a finger and slowly crush them to death. I have always assumed this was because their native hosts [deer] don't or can't swat them, but I may be wrong." A few months later,

Art emailed me to say that he was just back from New Zealand, where he experienced "the horrific blackflies on the west coast of the South Island (they call them 'sandflies' there). They too were dumb as planks, probably because they evolved without mammals to hit them." (New Zealand had no large mammals until humans arrived in the late 1200s.) "Those moas must have had a stink of a time before the Maori finished them off."

You would have no trouble swatting a bat fly or a louse fly, but that would require you first to encounter one, which is an unlikely prospect, given their obscure lifestyles. I've seen bat flies on only one or two occasions, back when I was studying bats in the wild.

Many bat flies cannot fly, and some are wingless. After all, why bother flying when your winged hosts can do the flying for you. These weird insects can be identified as flies only after detailed examination, as Steve Marshall explains:

"Males and newly emerged females are normal-looking fully winged bat parasites, but [a few] females undergo extraordinary transformations once they find their hosts. Upon arrival on an appropriate host the female burrows almost entirely under the bat's skin, loses her wings and legs, and undergoes extensive bloating of the abdomen so that it envelops the head and thorax. Once encysted under the bat's skin, the female fly becomes little more than a bloodsucking bag that contains a developing larva; it is hardly recognizable as an insect, let alone a fly."

Only the posterior tip of the female's body protrudes from the bat's skin, allowing the "fly" to push out a single, large larva that has matured in her uterus. The fat grub drops off and pupates almost immediately after hitting the ground, hatching out some weeks later to complete the life cycle. It's a testament to the strange twists of evolution that a young fly, lavishly nurtured by its mother, should ultimately spend what would appear to be a very dull adulthood buried headfirst in the flesh of its host.

Most of the closely related louse flies live among the feathers of birds. One, however, the diminutive 4-to-6-millimeter-long (one fifth of an inch) sheep ked, is wingless, and spends its entire life cycle nestled in cozy fleece niches, sipping blood from its hoofed host. Like the wingless bat flies, a mother ked invests heavily in her single larva, which feeds from a "milk" gland in the female's uterus. These flies are found virtually worldwide, having benefited from the globalization of sheep raising. Insecticides and quarantine measures appear to have been the cause of their recent quiet disappearance from much of the United States and Canada.

Hidden Benefits

Our antipathy toward blood-seeking flies is understandable. Nevertheless, assuming these aerial phlebotomists are here to stay, we may take heart in a number of benefits they provide in addition to the incalculable one of being food for countless other animal species, as both adults and larvae.

One of the subtlest benefits of the bloodsucking fly lifestyle comes from the fact that blood is a reservoir of information that we can use—to solve crimes, for example. Flies that draw blood and flies that sup on carrion ingest traces of DNA from the animals whose tissues they feed on. Human DNA profiling from blood collected from engorged mosquitoes is feasible for up to three and a half days after a blood meal. While any trace of dinosaur DNA has long dissipated from the guts of sated mosquitoes or biting midges preserved in amber, modern human DNA recovered from a mosquito can be matched to blood samples from a murder suspect. Thus, mosquitoes found at a crime scene, dead or alive, can harbor valuable forensic evidence. We will dive into the lively field of forensic entomology in chapter 11.

Fly-imbibed blood is also helpful to wildlife biologists, who have been able to detect the presence of pathogens threatening to wildlife. From a sample of 498 flies collected at Taï National Park, in Côte d'Ivoire, using mosquito traps and traps baited with meat, 156 of the flies (31 percent) contained mammalian DNA. Although just 8 of these insects contained DNA samples suitable for amplification and sequencing, the researchers were able to detect and identify ten kinds of primates and rodents. The presence of adenoviruses (common viruses that cause a range of illnesses in us, most notably coldlike symptoms) in most of the samples indicates that these pathogens are quite widespread in the region. One sample contained what is likely a new rodent adenovirus. This approach could be used for early detection of pathogens that threaten mass wildlife mortality in pristine areas. Currently, the methods used here are not cost-efficient, but the advance of DNA techniques could change that.

Which brings me to one of the least recognized benefits of blood-feeding flies.

Flies prevent habitat and biodiversity loss by curbing the human presence in ecologically sensitive areas. The tsetse fly is a notable example.* The colonial era and the subsequent ups and downs of cattle farming in Africa are entwined with the tsetse. Human sleeping sickness is transmitted by the tsetse, along with nagana, which afflicts cattle (but less so indigenous species, which have evolved resistance). Conservationists sometimes call the fly "the best game warden in Africa." So vital is the tsetse considered to the preservation of southern African biodiversity that conservationists in South Africa have challenged as unconstitutional a proposal to (try to) eradicate these flies.

Certainly the specter of harmful diseases is nothing, by itself,

*There are actually 23 known species of tsetse fly.

to celebrate. Relatively benign biting flies can also be effective conservationists. Human population densities and the numbers of roads are markedly lower in areas of Scotland where the local biting midge thrives in huge numbers. Another biting midge has kept developers out of mangrove swamps in the New World tropics where it breeds.

I find it oddly satisfying that a lowly biting fly can wield such influence over the fate of an ecosystem. We also don't usually think of insects as being at the top of a food chain, but consider that a biting fly may alight and dine on a wolf or a tiger. Perusing Thomas Marent's beautiful 2006 pictorial book *Rainforest*, I turned to a close-up of a black caiman on the prowl, just the eyes and nostrils protruding from the water. A closer look revealed that the toothy reptilian predator is also serving as prey. Five mosquitoes, distended bellies glowing red, perch on the caiman's eyelids, from which they have just taken their fill.

Given blood's omnipresence and its nutritive punch, little wonder that insects, and flies in particular, have found ways to plunder this resource. But blood is not the only lurid foodstuff available to the opportunistic six-legged diner. Animals also excrete, and they die. In doing so, they leave behind a banquet of goodies "ripe" for the taking.

Chapter 6

Waste Disposers and Recyclers

What did one fly say to the other?
"Pardon me, is this stool taken?"

If there is one aspect of flies to rival their annoying and sometimes dangerous biting habits, it is their association with filth and decay. We may well ask how any creature could find irresistible something that to our senses is so disgusting as excrement or rotting flesh. Our own disgust probably evolved as a mechanism to minimize contact with potential reservoirs of pathogens. But the inevitable waste products of life also make for a safe and cozy shelter that conveniently contains valuable nutrients and calories, so we need hardly be surprised that so many animals (and plants) have evolved to use them.

I find it ironic that we so often demean creatures as lowly for their affinity for dung and decomposition. Disgust is one thing, but disdain? How many of us have paused to reflect on the vital role flies play in helping to keep the world clean? Consider that the average American produces about 11,400 kilograms (25,000 pounds) of poop in a lifetime. That's about the weight of three adult hippos. Multiply by a few billion humans for a global extrapolation of about 1.3 trillion kilograms (1.5 billion tons) of human dung per year.* And that's not counting all the nonhuman poop deposits. I don't know about you, but my first thought is: "Thank goodness it's organic waste that will break down." And who's breaking it down? Lots of little organisms. Were it not for the flies (and beetles), the world would soon be awash with organic waste, and we'd be living in a much fouler and more pestilential place. Feeding on excrement is not just an extremely good strategy in the evolutionary sense, but a gift to anyone who values hygiene.

The same goes for dead bodies. Where vertebrate scavengers like dogs and vultures are scarce, flies typically make up most of the invertebrates on and in a carcass, and they can consume over half of it. It bears noting, in Canadian fly authority Stephen Marshall's words, that "every living thing, including you and me, will eventually die and decompose, and the odds are good that our reintegration with the primordial ooze will be speeded along by microbe-munching maggots."

Will Work for Shit

When an acquaintance of mine learned that I was writing a book about flies, he handed me a button with a cartoon drawing of a fly and the words: WILL WORK FOR SHIT. Oblique references to minimum wages aside, coprophagous (poop-eating) insects do pursue their craft with admirable zest.

Like it or not, if you spend any time outdoors, you are bound to encounter flies on feces. Shit happens, and no matter how

*25,000 pounds per lifetime × 7 billion humans = 175 trillion pounds ÷ 2,000 pounds per ton = 87.5 billion tons for all current human lifetimes = about 1.5 billion tons per year (given a global average human life span of 60 years).

hygienic the community one lives in, it seems there are always one or two rogues who don't clean up after their dogs. Even when they do, there are always flies alert for opportunities. A couple of years ago, when I was living in a condominium in south Florida, my neighbor's large dog Paddy defecated on a patch of lawn out back. Bridget is a conscientious dog guardian and she had poop bags at the ready, but she politely waited until we were finished chatting before attending to it. Four minutes had elapsed by the time we both ventured onto the lawn to find Paddy's latest attempt to fortify the local ecology. It was easily found, but it didn't look like the brown pile one might expect. It shimmered blue-black, plastered with the metallic bodies of a hundred or more bluebottle flies. They all buzzed off as Bridget descended with the bag, and we gasped at the speed with which these insects had located their prize.

The prevalence of solid waste is proportional to the prevalence of creatures that produce it. Thus it would not appear to be a scarce resource. In the Anthropocene, however, two factors may be reducing the niches of scat-seekers. First, humans go to great lengths to sequester our fecal waste and to process it out of the reach of creatures that might otherwise consume it. Second, our growing global ecological footprint is causing biodiversity loss, and that means fewer producers of the food that scat-seekers seek.

To get a feel for the efficiency with which flies process dung, it helps to have some numbers. A pair of biologists from Chengdu Institute of Biology and Nanjing University placed 180 artificially formed pats of fresh yak dung (17.6 centimeters diameter x 4 centimeters high = 7 inches x 1.6 inches) in an alpine meadow in western China. Each pat weighed about 1 kilogram (2.2 pounds), with a dry weight of ¼ kilogram. Using fine mesh coverings to block fly access to 45 of the pats, they tracked the progression of

Hundreds of small dung flies feast from a single pile of fresh horse manure in Maryland.

(PHOTO BY THE AUTHOR)

dung weights over 32 days. With some clever mesh work, they also measured the effects of dung beetles (with flies excluded) on another 45 dung pats, and the combined effects of both flies and beetles on another 45. A fourth set of pats excluded access by both flies and beetles. To the beetles' credit, they outscored the flies, reducing their exclusive pats by two thirds during the monthlong study, compared with a half for the flies. Dry weight of the set of 45 pats from which both flies and beetles were excluded dropped by just a tenth.

I've seen dung beetles in action in both Africa and Asia, and have witnessed firsthand the impressive speed with which they will descend on fresh dung. I once had to take an urgent "bathroom" break while monitoring a mist net set up to catch bats in a riverine forest in Kruger Park, South Africa. To my surprise, several large dung beetles had arrived and were already negotiating the apportionment of my dung before I had a chance to bury it. Elsewhere, on one of the few paved roads, I noticed that a large pile of elephant dung had been reduced to a scattered mat three hours later. This pattern of early arrival and dung removal explains why the beetles in the Chengdu study had reduced their yak pats by almost two thirds in just the first 2 days. By contrast, the flies' yak pats underwent a steady decline over 11 days.

Houseflies are no slouches at dung decomposition, and their reproductive output helps. A typical housefly lays 2 to 7 batches of 50 to 100 small, elongated eggs (25 laid end-to-end equal an inch). But the total can vary; the record output for one fly's lifetime is 21 batches with a grand total of 2,387 eggs. Eggs hatch in 6 to 30 hours. After a half-ton pile of stable manure was exposed to the egg-laying activities of female flies for 4 days, random sampling estimated that the pile contained about 400,000 fly larvae. Larvae pass through three molts, attaining a maximum of half an inch in length, then move to drier locales to pupate. Different conditions affect the duration of each stage: larval, 3 to 14 days; pupal, 3 to 10 days. An adult's average life span is 30 days, with a maximum of 70.

Being a poop-eater doesn't necessarily mean you're abundant. Many flies, including some that go for doo-doo, are rare and obscure. Choosy connoisseurs, they select only the dung of a specific host. In Australia, for instance, some flies occur only on wombat scat, and a wingless bat fly has a taste only for the feces of New Zealand's endemic short-tailed bat. One bat-dung-eating fly has only ever been found in a single cave in Kenya.

This brings to mind a true story involving a fly researcher who also selected only the dung of a specific host—his colleagues—and it nearly led to a fistfight. The dedicated dipterist was struggling to find enough dung to conduct his studies in the tropics, where excrement disappears at a very fast pace. On a collecting expedition, instead of asking fellow dipterists to contribute to the cause, he would surreptitiously track individuals into the bush where they would be doing their job. When the coast was clear, he would collect the feces and then use them in his flytraps. This scheme worked for a few days until one of the senior members of the group figured it out and was so offended that he launched a screaming, in-your-face tirade, demanding an end to the shenanigans, adding, "If you want my shit, you can ask for it." The entomologist who confided this to me added that it puts a whole new spin on giving somebody shit.

An Animated Sparrow

Excrement isn't the only thing we can be grateful to flies for cleaning up. Like dung, dead bodies present opportunities for dinner, and flies are among the main beneficiaries. If you have inspected the ripe carcass of a dead creature, you have most likely beheld dozens or more squirming maggots busily converting the carcass into fly biomass.

I once noticed a dead bird on the wooden deck behind my townhouse. A neighbor's bird feeder had been attracting numerous house sparrows, and it was a female, probably from this flock, who had somehow met her end on our deck. Over the years I lived there, the occasional bird collided with our rear windows, in spite of the silhouettes of falcons we taped to the panes.

A familiar odor of rotting flesh indicated that the bird had been dead at least a day, yet to my amazement, the body was moving in a manner suggesting the creature was still breathing. It was not until I gently lifted the bird onto a plastic bag that I discovered the movements were generated by the foraging of maggots beneath the skin and feathers. I was reminded of a time-lapse film showing the rapid decay and re-earthing of a dead mouse by waves of maggots sweeping back and forth beneath the animal's pelt.

There is order to this process. A sequence of decompositional stages—fresh, bloated, decay, post-decay, and remains—is accompanied by a succession of adult flies and then maggots at different stages of development. Such is the predictability by which certain species colonize a corpse at different stages of deterioration that flies have become an important aid to resolving time of death (see chapter 11).

The maggots animating the bird carcass on my deck were probably blowflies or flesh flies. Blowflies number about 1,100 species, most of which are associated with carrion. Named for an old sense of the verb *blow*, meaning "to lay eggs," blowflies are noted for their brilliant metallic colors. They include the well-known bluebottle and greenbottle flies, although they also come in gleaming copper and purple. The gorgeous appearance of these adult flies belies the morbid milieu of their childhoods.

Flesh flies comprise about 2,500 species. They don't wear the flashy attire of their blowfly cousins, instead typically clad in more muted grays or browns, with handsome pinstripes adorning their muscular thoraxes. Another distinction between flesh flies and blowflies is that flesh flies do not lay eggs; they give birth to live young, which begin feeding immediately. Despite their name, most species of flesh flies do not normally colonize dead creatures, at least not vertebrates.

The Necrobiome

Their cadaverous habits make blowflies and flesh flies members of the *necrobiome*, the community of interacting organisms whose livelihoods are tied to the process of animal decomposition. The necrophagous (carcass-eating) flies are predominant early in succession, when soft tissues are easily liquefied to allow access to the abundant nutrients in the carrion. It's an intensely competitive arena, and the larval stages are designed for rapid progression toward pupation. High metabolic rates are further fueled by heat generated by the feverish activity of hundreds or thousands of maggots whose stretchy skin reduces the need for molts to just three. Fat maggots and relatively defenseless pupae soon become attractive targets for predatory beetles, ants, wasps, hornets, and parasitic wasps. Later in the cascade, other maggots and beetle larvae take the reins of rot, specializing on the drier portions of carcasses, like tendons and mummified skin.

One way to study the invertebrate necrobiome is to place a recently dead body in a secluded location, inside a semipermeable barrier to keep out scavengers, then collect colonizers over time. That's what a team of Brazilian researchers did in a forest in southeastern Brazil. Over the course of four seasons, they placed 8 freshly killed rodents (4 rats, 4 mice) inside iron cages in sunlit and in shady areas, a total of 32 rodents. The scientists returned regularly to collect and identify insects that had colonized the carcasses.

In all, 6,514 arthropods colonized the decomposing bodies (820 adults and 5,694 immatures). They were represented by four major groups: flies, hymenoptera (mostly ants), daddy longlegs, and beetles. The flies dominated the scene, making up over 95 percent of the total. They were also by far the most diverse, numbering 44 species in 15 families, compared with 4 kinds of ants, 2 wasps, 1 bee, 1 beetle, and 1 daddy longlegs. In the competitive world of putrefaction, these mini beasts work fast. In the warmer spring and summer seasons, it took less than six days for a carcass to decompose to the point that it was no longer visited by arthropods.

A Venezuelan study found a similar predominance of flies. Four days after they placed 4 kilograms (10 pounds) of cow viscera (liver and lungs) in an exposed plastic container in an urban location, two researchers from the Universidad de Carabobo collected 1,046 adult insects from the rotting offal. Almost all (97 percent) were flies of 11 species; the remaining 3 percent were 3 kinds of beetle. Curiously, the investigators counted only adult insects, but photographs indicate that, as usual, larvae far outnumbered the adults who spawned them.

While hundreds or thousands of hungry maggots eating a rotting carcass is a competitive feeding situation, there is also cooperation. The combined digestive enzymes of a team of maggots and the collective effects of their mouth hooks gnawing on the flesh causes rotting meat to break down faster. The maggot mass also generates impressive heat, up to 30°C (54°F) above ambient temperatures—a remarkable feat for "cold-blooded" creatures. Heat is believed to be generated by the friction of all those wriggling grubs, and perhaps also metabolic and microbial activity.

Synergy notwithstanding, a dead body is a risky place to spend one's adolescence; larger scavengers may include maggots in their diets, so it pays to get in and out as soon as possible. This probably explains why egg-laden flies tend to cluster their eggs where other females have done so, and why some blowflies' eggs are scented with pheromones attractive to other flies. The higher temperatures in the feeding mass have their own benefits: accelerated growth rates that reduce exposure to predators and parasitoids, and protection against sudden drops in surrounding temperature.

Inevitably, some larval masses get plundered before they are ready to disperse—another of flies' ecological benefits. Bob Armstrong explained to me that juvenile birds, naive in the craft of exploiting more challenging food sources, might especially benefit. Bob, who is based in Juneau, Alaska, has worked as a fishery biologist and research supervisor for the Alaska Department of Fish and Game and as an associate professor at the University of Alaska, where he taught courses in fisheries and ornithology. In the post-spawning glut of salmon carcasses, many seethe with maggots. Bob has been watching and filming them, and so far he's identified 12 species of birds gorging on the maggots, including sparrows, thrushes, wrens, and ducks. This is occurring at a time when many young birds have recently fledged, and as far as Bob can tell, most of the birds are juveniles. "I suspect they are fairly inept at obtaining food at this age," Bob told me, "and the maggots provide an easy-to-capture and highly nutritious food source." Bob sent me a short video of juvenile and adult northwestern crows feeding on pearly-white maggots, scooping up several at a time with their large beaks. The birds' table manners are far from neat, and many maggots either get dropped or left behind to continue their own recycling duties.

The efficiency with which carrion-loving flies colonize dead beasts was made plain to me when I found a dead lizard on a footpath behind my former apartment in south Florida. I could see that the creature had been deceased for some hours, with a gaping hole in its side perhaps inflicted by a bird. Close to 100 bluebottle flies swarmed over the 5-inch carcass. A few smaller flies also darted around excitedly. When I approached to within four feet, like nervous vultures the flies took flight in a loud buzz. I stopped, and within ten seconds, they were alighting again, by the dozen.

I dragged the lizard about two feet off the path, depositing it beneath a shrub where the necrobiome could do its work relatively undisturbed. Then I waited to see how soon the flies would return to their meal. I naively expected them to shift their attention wholly to the lizard's new location. Indeed, a minute later there was a growing cluster of flies on the relocated carcass. I was not too surprised to notice a second cluster of flies—about fifty of them—returning to the wet spot on the footpath left by the deliquescing carcass. What I didn't anticipate was an orderly row of flies alighting along the invisible line across which the carcass had been dragged.

This scene put me in awe of the finely tuned sensitivity these flies have for the odors and tastes of decomposing flesh. I am impressed not just by the fidelity of their detection systems, but by how promptly they act on it. This happened a few months before Paddy's turd attracted all those bluebottle flies nearby, and I wonder if some of the dog's flies were descendants of the lizard's ones. As an observer of animals and nature in all its manifestations, I have looked upon many scenes of death and defecation. I cringe at the tragic fate of so many animals killed on the road, and I have pulled many into roadside ditches where scavengers can work in relative safety. Almost without fail, flies are present, and they are a reminder of the vital role insects play in the efficient cycling of nutrients through ecosystems.

Composters

Compared with a decomposing corpse or a fresh turd, a pile of compost is less disagreeable to our senses, and positively irresistible to one particularly valuable fly. If you've kept a composter in your yard, chances are good that you've encountered black soldier flies (*Hermetia illucens*), a species noted for being a champion composter (see the image in the photo insert). Their larvae were the most prevalent of the larger detritovores that colonized the composters I kept in Maryland and also in Florida a thousand miles to the south. The only other similar-size maggot I remember seeing in my composters were the rat-tailed maggots destined to become hoverflies, named for a long snorkel at the tail end that can be telescoped to an incredible 15 centimeters (5.9 inches), allowing them to breathe while churning through wet muck.

Within a few months of establishing my composters, I was seeing the fairly large (nearly half-inch), somewhat flattened black soldier fly maggots squirming amid the decaying mass inside. Soon some were inching up the sides and venturing out to find a suitable pupation site. (In Florida I rescued many whose wanderings had taken them into the swimming pool, where some had drowned but others had probably withstood several hours in the water.) By midsummer I began to see 1.6-centimeter (5⁄8-inch) slender black flies perched on the edge of my composter, and I soon learned what they were.

The soldier fly family to which the black soldier fly belongs comprises 2,800 known species worldwide, most of whose larvae feed on decaying vegetable matter before making themselves useful again as flower pollinators in adulthood. Originating in the New World, these handsome flies, with a pair of longish antennae pointing forward, have spread to all continents and are now virtually cosmopolitan.

A mother black soldier fly bears about 600 babies per litter. Each larva can eat a gram of compost—several times its own weight—per day. No wonder these insects can fetch $330 per ton on the feed market. The use of black soldier larvae to manage organic waste by converting it into valuable protein-rich animal-feed ingredients and biofuel has developed into a global industry. In

Cape Town, South Africa, a branch of a global company named AgriProtein has generated a two-for-one strategy, using black soldier larvae to recycle city waste, then as protein-rich feed before they pupate. While federal regulations so far prohibit their use as livestock feed in the United States, it's a business opportunity that has not been lost on North American entrepreneurs. Enterra Feed Corporation, founded in 2007 and based in British Columbia, is reportedly building a 180,000-square-foot facility in southern Alberta, where it plans to grow billions of black soldier flies on the leavings from fruit, vegetables, and other foods. Once fattened up, the larvae will be incorporated into feed eaten by chickens, fish, and other livestock. South of the border, the American company EnviroFlight, launched in 2009, currently sells four product lines: EnviroBug (whole oven-dried larvae), EnviroMeal (oven-dried larvae pulverized into powder), EnviroOil (oil mechanically pressed out of the dry larvae), and EnviroFrass (production leftovers, including larvae waste, exoskeletons, and remaining feed ingredients). The company's target markets are fertilizer and feed for poultry, aquaculture, pets, exotic animals (in zoos, for example), and young livestock (pending legalization).

Feeding animals to billions of humans is inherently unsustainable (and inhumane from the animals' standpoint), but feeding the animals on soldier-fly spawn is more sustainable, because "we can grow them without arable land," notes Victoria Leung, Enterra's VP of operations. Furthermore, "no added water is required as all the water the insects need comes from the recycled fruits and vegetables in their diet." These are significant advantages, considering that conventional animal agriculture takes up about 60 to 80 percent of all agricultural land* and consumes well over half of the

*According to globalagriculture.org, pasture and arable land dedicated to the production of livestock feed represent almost 80 percent of all agricultural

freshwater that humans use.* Enterra touts its product as "a highly efficient, low-impact source of nutrients compared to resource-intensive alternatives like beef, pork, chicken, fish meal, soymeal, coconut oil and palm kernel oil."

I reached out to Enterra and EnviroFlight, hoping to get some further clarification on aspects of the industry not available on their company websites. In addition to the basics of production volume, conversion rates, and product distribution, I was interested to know what proportion of larvae are allowed to metamorphose into adult flies for breeding stock and what sort of environment the breeder flies require for breeding. I was also curious to uncover the process of converting live maggots to dead ones. Are they just tossed wholesale into the ovens and roasted alive? Are they chilled and frozen first? Something else? I explained that my book would include the debate around the possibility of insects being sentient.

I never found out. The Enterra representative called and invited me to email him my questions, but he subsequently didn't respond to my follow-up emails and voicemail. The EnviroFlight representative also invited me to send questions, then having read them replied that she was "not able to share any information about our process that is considered intellectual property." Such reticence probably bodes ill for the larvae. If maggots turn out to

land: https://www.globalagriculture.org/report-topics/meat-and-animal-feed.html (accessed May 5, 2020). According to the United Nations Food and Agriculture Organization, more than a quarter of all land is taken up by livestock grazing, and a third of all arable land is used to grow feed crops.

*Cutting consumption of animal products in half would reduce the United States' dietary requirements of water by 37 percent, according to a recent report from *National Geographic*: "Thirsty Food: Fueling Agriculture to Fuel Humans," https://www.nationalgeographic.com/environment/freshwater/food (accessed May 5, 2020).

be sentient (see chapter 3), this industry might need some public relations magic in its future.

In a last-ditch effort to glean pertinent information on the fate of industrially produced black-soldier-fly larvae, I emailed Jeffery Tomberlin, entomologist at Texas A&M University and CEO of EVO Conversion Systems, LLC, a consortium covering all aspects of the black-soldier-fly industry.

"Are the grubs, say, chilled or frozen first, or are they tossed live into the roasting ovens?" I asked.

Tomberlin replied that "it will vary depending on the company—so 'all of the above' would be appropriate as a response."

It appears, then, that being roasted alive is a very real possibility for at least some of these larvae. A video linked from EVO's website includes a scene of apparently live larvae being poured into boiling oil at a Chinese facility. Surely, death for such a small organism would occur within just a second under that circumstance, but it all gives me an uneasy feeling. Perhaps it is some consolation to mass-produced soldier flies that several million will be allowed to mature into breeding adults, according to a 2018 article in the Calgary press.

Black soldier flies are notably versatile. At the other end of the livestock food chain, they have been used extensively in poultry-breeding facilities to compost manure. And their sideline attraction to decomposing bodies has made them useful accessories to crime solving.

Black soldier flies are not the only flies being used or considered as feed for animals being reared for human consumption. The common housefly is under research scrutiny as livestock feed, even as a replacement for fish meal in aquaculture operations. The authors of a 2017 paper describe houseflies as a potential source of livestock feed with dual benefits: first, the larvae can be raised on livestock manure, thereby reducing waste disposal.* Then the resulting insect biomass can be used as a protein-rich animal feed. In the enormous aquaculture sector, which today accounts for nearly half of all fish consumption by humans, housefly larvae could ease some of the heavy burden that aquaculture places on wild fish populations. In 2010 it was calculated that 73 percent of fish meal and 71 percent of fish oil produced were consumed by aquaculture operations.

Preparing for a Tiny Meal

Flies' attraction to compost and sundry other comestibles may have given rise to a curious behavior. If you take the slightest interest in insects, you will almost certainly have noticed how a fly will land on your arm or tabletop, then rub its forelegs together like a tiny gourmand preparing for a meal. So purposeful does this behavior look, and so reminiscent of a human diner, that I half expect the fly to reach for a napkin and snap it open with a flourish before tucking it in.

Why do they do it? Humans have pondered that question for ages. According to Joanne Lauck Hobbs, whose 1998 book *The Voice of the Infinite in the Small* explores our relationships with insects, the California Mission Indians believe the fly rubs its "hands" together in a supplication, to beg forgiveness for speaking harsh words from which people died. The Luiseño peoples of what is now Southern California have an ancient myth in which a fly created fire for the mourning ceremony of a dying leader by rubbing a stick between his hands. Blue Fly (bluebottle fly)

---

*Livestock produce over 335 million tons of manure (as dry weight) per year, too much to safely apply to field crops as fertilizer (Hussein et al. 2017).

twirled the stick for so long that he couldn't stop, even to the present day.

Here's a modern scientific interpretation, courtesy of science journalist Nicholas DeMarino:

"Flies rub their limbs together to clean them. This may seem counterintuitive given these insects' seemingly insatiable lust for filth and grime, but grooming is actually one of their primary activities. It gets rid of physical and chemical detritus and clears up their smell receptors—all of which is important for flying, finding food, courting mates and just about everything that a fly does."

I think of it as a fly's version of cleansing one's palate.

Unfortunately for us, flies aren't washing their hands with water or soap, and their diverse tastes—which include peaches, pilaf, and poo—have gained them a well-earned reputation as vectors for the spread of undesirable microbes.

Flies' mobility doesn't help. Studies of marked flies have shown they are always on the move; one was found 21 kilometers (13 miles) away a few days after its release. I have seen stray houseflies and fruit flies inside aircraft zooming over oceans at 600 miles per hour, and surely some of these jet-setting stowaways successfully deplane at their destinations. It makes me wonder what a fly might experience upon arriving in a distant land—say, in Nairobi from New York. All those foreign smells! Can a fly feel bewildered?

Needless to say, those stowaway flies have stowaway microbes on them. Research suggests that flies are about twice as filthy as cockroaches. In the 1940s, American entomologists David T. Fullaway and Noel L. H. Krauss of the Hawaii Board of Agriculture and Forestry measured the bacterial loads of houseflies by having their subjects walk across a nutrient-rich gelatin-covered plate, then studying the white tracks, blooms of bacteria, that appeared a few days later. Fullaway and Krauss calculated that the average housefly carries 1,250,000 bacteria on it. One notably unhygienic individual in their study carried 6,600,000.

John Wallace, a biology professor at Millersville University, told me that a colleague, John Diehl with the US Department of Agriculture, used a similar method to demonstrate the contaminating effect of flies at roadside food stands in Guatemala. These makeshift vending facilities were set up at the many traffic stoppage points where cars are checked for contraband (and often sprayed for insect "pests.") Hungry flies were drawn in vast numbers to an array of tacos, enchiladas, cakes, and pies. The young scientists would scatter white flour around nearby latrines and wait. Within a short time, tiny fly footprints would appear on the displayed foods. The ploy proved effective in visibly connecting the flies' movements between toilets and tacos, and the locals took measures to improve the situation by covering their food items and using passive control measures, such as fly strips, to reduce cross-contamination.

Despite the evidence fingerprinting flies as accomplished filth vectors, a 2014 Orkin Pest Control survey found that 61 percent of respondents said that they would quit their meal after a cockroach touched it, compared with only 3 percent who said they would stop eating after a fly came in contact with the food. I believe it is cockroaches' larger size and their tendency to scurry (not fly) that inflames our prejudice against them.

What I find remarkable is that flies seem incapable of becoming sickened by their diets. Just think how careful we humans are taught to be regarding hygiene—with exhortations to wash our hands after every bathroom visit and before a meal—even before coronavirus came along. Millions of cases of food-borne illness go reported (and unreported) yearly. With all those headlines about the latest *E. coli* outbreak from a rogue batch of spinach or tomatoes—contaminated by animal manure, since vegetables don't have the colons for which *E. coli* is named—we may marvel at how well adapted flies are to their fetid food choices. Does a fly or a maggot never even get a stomachache?

We do know that flies are well adapted to the masses of bacteria present on dung and the dead. There is significant evidence, for instance, that the bacteria in or on a corpse don't merely fail to impede larval development and pupation; they promote it. Two ways this can happen are by larvae feeding directly on the bacteria, or the bacteria freeing nutrients for the larvae's consumption. While the maggots munch away, they also secrete or excrete a stew of antimicrobial molecules that can kill certain bacteria.

While supping on rotting corpses is appropriately disgusting to us, we might reserve a little gratitude for the vital role that flies play in cleaning up unpleasant messes. Sadly, however, that's rarely the case. I've more than once heard someone express disgust at a fly's habit of "vomiting" on a surface to liquefy any edible residue there. This seems biased to me. For one thing, it isn't really vomiting. We're not talking about the evacuation of previously ingested food, nor, surely, is it accompanied by any feeling of nausea. I'm not even sure it's any more disgusting or inherently less natural than our own manner of dousing our food with saliva *after* it enters the mouth.

For those whose sensibilities remain wholly offended, consider that a disgusting lifestyle may not be destined to remain so. Hoverflies might have the most aesthetically contrasting life-cycle transition from immature to adult. As maggots, they commonly dwell in the organic sludge found in sewage-treatment lagoons. Then, as nature's reward for such an unsavory adolescence, they emerge as adults to hover daintily on gossamer wings while sipping nectar from pretty flowers. Any notion of redemption here is, of course, anthropomorphic hogwash; to the extent that hover-fly maggots may have aesthetic awareness, a vat of human sewage is surely as delectable to them as a bowl of fresh three-bean chili is to us.

And thank goodness for that. I shudder to think what this planet would look and smell like if we didn't have flies around to clean up the messes other creatures inevitably leave behind when they poop or perish.

Chapter 7

Botanists

> Insects . . . are also our worst enemies but this fact should not be proclaimed from the housetops without at the same time crying much louder of the benefits they give us—flowers, fruits, vegetables, clothes, food, pure air, beauty.
>
> —CHARLES HOWARD CURRAN

Our tendency to associate flies with filth and decay overlooks a much grander association: that of flies with flowering plants. With the possible exception of the beneficial bacteria that live inside animals' bodies (including ours), the pact between flowering plants and their pollinators is the grandest mutualism on our planet. Mutualisms—in which two or more organisms benefit by associating with one another—provide some enchanting examples of nature's gift for opportunism. The enormously successful mutualism between plants and their pollinators has, given the luxury of eons, led to extravagant specializations between flies and the plants that employ their pollen-moving services. So far, scientists have largely neglected the possibility that pleasurable feelings may play an integral role in the insect pollinator mutualism, but the emerging evidence of insect cognition and sentience I outlined earlier in this book hints that "the botany of desire," to use Michael Pollan's fitting phrase, reaches far into nature's niches.

Many flies feed on plant tissues, and some are classified by us as serious pests, but how many of us stop to consider flies' hugely important role as pollinators? Of the 150 described families of Diptera, at least 71 include flies that feed at flowers as adults.

Although the world's phantasmagoria of brightly colored flowers brings pleasure to our senses, it evolved to attract mostly insects, who sip sweet nectar in exchange for a pollen courier service. Dusted with tiny pollen grains, the insect flies to another flower, where its efforts to imbibe the next round of nectar inevitably result in some grains being left behind and new ones picked up. It's a "fly food for plant sex" exchange.

About 218,000 of the world's 250,000 flowering plants, including 80 percent of species that humans define as food plants, rely on pollinators. Less than 7 percent of these food plants are pollinated solely by wind or water, less than 4 percent by birds, and less than 2 percent by bats, leaving close to 90 percent pollinated by insects.* So important is pollination that if it were the only benefit of insects, it would be enough to rank them among the most critically important of all animal groups. David MacNeal puts it well in his 2017 book *Bugged*: "Bugs are nearly as necessary to humans as breathing."

To place a dollar value on it, the commercial worth of food-plant pollination by insects is nearly a quarter of a trillion dollars globally.† Swedish entomologist Anne Sverdrup-Thygeson

---

*These percentages derive somewhat loosely from the following figures: Of approximately 13,500 genera (most of which represent many species) of flowering plants on Earth today, about 874 are solely wind- or water-pollinated species, about 500 are bird-pollinated, and some 250 include bat-pollinated plants. The remaining 11,900 or so genera are pollinated predominantly by insects.

†Nicola Gallai, an economist at the University of Montpellier, calculated in 2008 that the worldwide economic pollination value of insects averages $216 billion per year. See MacNeal 2017.

provides a more recent (2019) estimate: $577 billion. But dollar values are decidedly anthropocentric. We might also acknowledge the incalculable benefits to all nonhuman organisms that make up healthy ecosystems by their interdependent activities.

Overlooked Pollinators

As pollinators, flies have been overshadowed by their cousins the bees. By most estimates (so far), the Hymenoptera (bees and wasps) outrank flies (Diptera), beetles (Coleoptera), and butterflies and moths (Lepidoptera) as pollinators of flowers. But all these groups are major contributors, and in many locations bees take a backseat. In arctic and alpine environments, for instance, where weather conditions suppress bee activity, flies are often the main pollinators. "Most people just notice the bees," Art Borkent told me on a Skype call from his home in British Columbia. "But if you look more closely, you'll find many more little insects among the flower heads. These make up the vast majority of the pollinators in these climates."

Case in point: a six-week study of pollinators of 19 species of flowering plants in the alpine meadows of Mercantour National Park in France, between May and July 2012. Nearly two thirds of pollinators were flies, and over half of the total were members of the dance-fly family. The conclusion: "Flies widely replace bees as main flower visitors at altitude."

Similar patterns occur at high latitude. The short arctic summers are abuzz with insects, and most of the ones visiting flowers are flies. During observations of flowering plants of five species at an arctic location in Nunavut, Canada, in July 2010, hoverflies and houseflies accounted for 95 percent of pollination opportunities. A 2016 study by European and Canadian scientists who hid sticky paper traps among flowers in northeastern Greenland found that a housefly relative (*Spilogona sanctipauli*)* was an überpollinator of arctic flowers, echoing the findings of a study from arctic Canada four decades earlier.

Some cold-weather flowers are active fly recruiters. They attract flies by providing a warm shelter that can be more than 5°C (9°F) above ambient temperature. This keeps the flies' flight muscles warm and allows them to commute at temperatures that would thwart most bees.

Other arctic flowers use a stinky odor ploy to draw their fly allies. Bob Armstrong sent me an exquisite video he filmed near Juneau, Alaska, where he had mounted his mini video camera a few inches from the burgundy blooms of a rice root lily—better known to locals as the outhouse lily or lady-on-the-pot. Dozens of bluebottle blowflies swarm over the flowers. As they muscle beneath the swollen anthers and their brilliant metallic blue abdomens glint in the sun, their hairy backs soon become thoroughly dusted by golden pollen. It's an opulent scene.

Elsewhere in the world, depending on where you are, the diversity of flies visiting flowers can rival or exceed that of bees and wasps, regardless of altitude. In Australasia there are almost twice as many types of flies visiting flowers as there are bees and wasps. The reverse is true for the Neotropics.

Might flies have even been the first pollinators? "It's an ancient mutualism," Art Borkent told me. "The fossil record of biting midges goes back to a hunk of 97-million-year-old New Jersey amber. That's exactly the time that the diversification of flowering plants started booming. Is the explosion of flowers and flies just a correlation? Perhaps. Or is it more than that: a massive

*So far, it has no common name. Let me propose arctic flower fly.

adaptive radiation in two groups of organisms benefiting from the presence of each other? Almost certainly!"

Whether or not flies did it first, improved insect monitoring techniques are illuminating the diversity and sheer volume of flies' pollination services. In a European study, 1,762 insects were collected from flowers on the Swiss Alpine foothills over a 24-hour period. The sample encompassed 316 species from 10 orders; they were found on a total of 94 plant species. More than half of the insects (974, or 55 percent) were flies, of which there were 130 different species, compared with 61 kinds of bees and wasps.

Using radar to track the distinctive aerial signature of marmalade hoverflies, Jason Chapman, an ecologist at the University of Exeter, estimates that up to 4 billion of these insects migrate hundreds of kilometers between southern England and Eastern Europe each year. They visit billions of flowers, importing between 3 and 8 billion hitchhiking pollen grains into southern England each spring, and exporting as many as 19 billion grains each fall. Meanwhile, their predaceous larvae devour 6 trillion aphids, weighing 6,350 tons. Their influence is also felt higher on the food chain, where their collective nutritional bulk of 35 million calories helps feed countless birds, mammals, reptiles, amphibians, and fish.

An unexpected pattern is emerging in the types of flies doing the pollinating. Until recently, it was thought that a particular family of flies, the flower flies or hoverflies (Syrphidae) performed the lion's share of plant sex assistance. But it turned out that most of the flies visiting flowers on the Swiss pre-Alps—two thirds of the fly sample in one study, 643 individuals belonging to 85 species—were not flower flies.

Studies like those described above don't make headlines, and the bee-as-sole-pollinator myth gets constantly reinforced in popular culture. The 2018 British nature TV program *Rooted* includes a series of brief vignettes of insects visiting flowers. As the cameras pan across sumptuous blossoms, there are more flies than any other insects, yet only bee and beetle pollinators are mentioned.

I experienced a similar pattern of fly neglect while visiting the Montreal Insectarium, which adjoins the much larger Botanical Garden. Despite its fairly small size, the Insectarium is allegedly the largest insect museum in North America. You wouldn't know it if you were a fly. In the lobby a backlit display had photos of 41 insects, and even though it includes a section on "star pollinators," not a fly could be found among the usual suspects: bees, wasps, and bumblebees. (As I type this in Toronto's international airport, a fruit fly is skittering energetically around my beer glass, as though egging me on.)

Not only are flies overlooked in favor of bees, but they are also mistaken for them. Many flower flies, especially, have evolved body shapes and black-and-yellow-striped abdomens to closely resemble their stinging cousins, thereby to scare off would-be predators without having to invest in expensive venom. So convincing is the mimicry (see the image in the photo insert) that many an entomology student has mistakenly included flower flies in their hymenoptera collections. Issue 57 (October 2016) of the North American Dipterists Society newsletter *Fly Times* includes a brief note and a photograph depicting a jar of honey whose label features an attractive photograph of a "bee" sitting on a pink flower. The company failed to notice that the insect is not a bee but a hoverfly mimic, from whose guts no honey shall ever issue. The prior issue of *Fly Times* includes a clip from a major newspaper lamenting the decline of bee populations due to a deadly bee virus. The accompanying photo is of a drone fly. As the entomologist F. Chris Thompson wryly notes, apparently the bees are getting so scarce that the newspaper couldn't find a photo of one. But sometimes the tables are turned. Directly below the drone fly,

Sparring male cactus flies from Mauritius.
(© STEPHEN MARSHALL)

This biting midge has "plugged into" a lacewing's wing vein for a meal of hemolymph (insect blood).
(© STEPHEN MARSHALL)

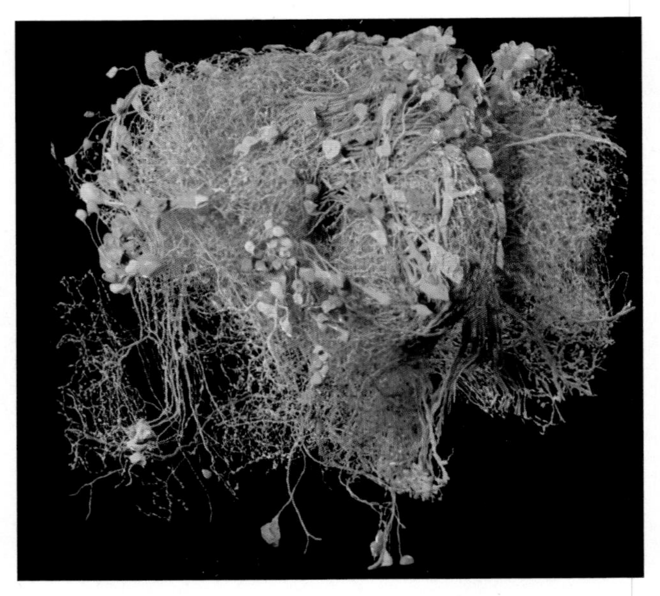

A detailed map of a portion of a fruit fly's brain containing some 25,000 neurons, among which can exist some 20 million interconnections.

(© JANELIA RESEARCH CAMPUS/FLYEM)

A male big-eyed fly, whose holoptic vision allows him to spot a female (or a foe) from almost any angle.

(© KATJA SCHULZ)

Robber flies can and do catch prey larger than themselves, but this one from Singapore is feeding on a tiny cousin.

(© VIN PSK PHOTOGRAPHY)

A mosquito from Ecuador with brushy adornments on her legs drills for a meal on someone's lip.

(© STEPHEN MARSHALL)

The calls of this amorous male túngara frog from Panama have attracted two frog-biting midges.
(© XIMENA E. BERNAL)

A phorid fly hovers above a colony of imported red fire ants in Canada, awaiting an opportunity to dart in and deposit an egg.
(© JOHN AND KENDRA ABBOTT)

A South African meganosed fly deploys her extremely long tongue to get nectar from a flower that coevolved with the fly.
(© ANTON PAUW)

Flies such as this bumblebee-mimicking drone fly photographed in Ontario are hugely important pollinators.
(© STEPHEN MARSHALL)

The (stingless) soldier fly from Ecuador, on the left, benefits by impersonating the wasp on the right from the same region.
(© STEPHEN MARSHALL)

By holding out their disruptively banded legs during flight, phantom crane flies probably confuse predators into attacking a leg, sparing the rest of the fly.
(© KAROLINA STUTZMAN)

A pair of snipe flies demonstrate the end-to-end mating position of many flies.

(PHOTO BY THE AUTHOR)

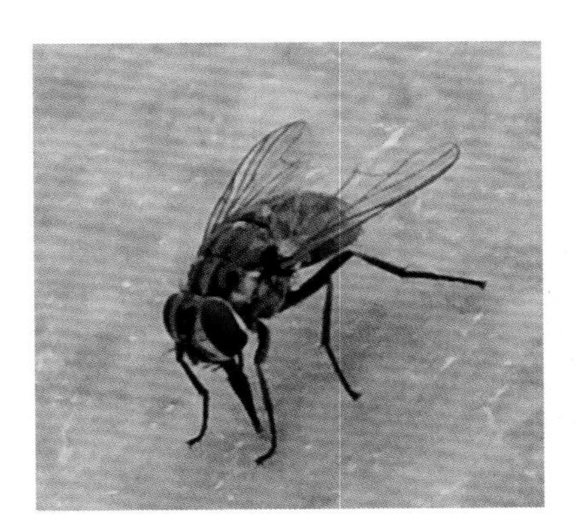

A stable fly in Ontario just before and just after filling up on the author's blood.

(PHOTOS BY THE AUTHOR)

Black soldier flies are champions at composting. The use of their larvae as high-protein food for animals and humans is a growing industry.

(© JOSEPH MOISAN-DE SERRES, QUEBEC MINISTRY OF AGRICULTURE, FISHERIES AND FOOD)

More than a dozen aptly named freeloader flies brazenly sponge up liquefying tissues of a honeybee caught by a lynx spider in Cuba.

(© STEPHEN MARSHALL)

Thompson has provided a cover image of an edition of William Golding's iconic novel *Lord of the Flies*, featuring an illustration of a large insect perched menacingly over a chubby schoolboy. The insect, with four wings and a stinger, is decidedly not a fly. As Aristotle noted nearly 2,400 years ago when he coined the name *diptera*, "No two-winged insect has a sting at the rear."

Delectables

Honey notwithstanding, chances are that some of your favorite foods grew thanks to the activities of flies. Many of our favorite fruits are at least partly pollinated by flower flies, including apple, pear, strawberry, mango, cherry, plum, apricot, raspberry, and blackberry. It was while researching this book that I learned that the early autumn pawpaws I used to eat while cycling along the Potomac River in Maryland are pollinated by flies. These delectable fruits, too perishable to be grown commercially, present a delicate flavor in a custard-like consistency. Flies also pollinate herbs and vegetables, including fennel, coriander, caraway, onions, parsley, and carrots. In all, more than 100 cultivated crops are regularly visited by flies and depend largely on fly pollination for abundant fruit set and seed production.

Chocolate is hardly a staple, but given its reputation as one of the world's most beloved edible substances, its revered place in human culture is secure. The cacao tree, from whose seed pods chocolate is made, also happens to be one of the world's most difficult plants to pollinate, and we may thank Allen Young, an American entomologist who spent several years studying this pollination system in Costa Rica, for shining a good deal of light on how it works. As a self-incompatible tree, cacao is unable to fertilize itself and therefore requires insects for successful pollination.

With the possible exception of one species of flower fly, only very small midges are known to pollinate the penny-size, whitish-pink flowers emerging directly from the cacao plant's trunk and lower branches. As Young explains in his 2007 book *The Chocolate Tree: A Natural History of Cacao*, the five petals of each flower point inward, leaving a tiny opening through which only a proportionately diminutive insect might squeeze. The midges themselves are lovingly described by Young as "so tiny that they resemble barely visible specks of dust zipping through the shady undergrowth of the cacao." The specimen that Mark "Doctor Bugs" Moffett photographed is a wispy creature with long, frail-looking legs and an elegant pair of long antennae, segmented like a string of pearls, curling back over a body cloaked with wide gossamer wings.

Human noses are unable to detect any aroma emanating from cacao flowers, but when Young and colleagues at the University of Wisconsin presented midges with cotton balls soaked in floral oils distilled from the flowers, the tiny flies flocked to the baits. Although cacao trees can produce prodigious numbers of flowers, even with the midges' assistance only a tiny fraction of these will ever grow into a mature cacao pod. The moist understory of decaying leaves below provides ideal habitat for the larvae. If you're interested to learn about this pollination system in more detail, I recommend Young's book.

The pollination of jackfruit is less celebrated but no less noteworthy. Jackfruit is quickly growing in popularity in the West, fostered in part by demand from Asian immigrants already familiar with its delights. Today it is cultivated in tropical and subtropical regions worldwide.

I made acquaintance with jackfruit while living in South Florida, and it is one of the most interesting and delectable fruits I have tried. I purchased one from a local market, then a second

Tiny midges are the only insects known to pollinate the complicated flower of the cacao tree, from whose fruit chocolate is made.

(© VINAYARAJ VR)

from a friend whose yard is home to several tropical fruit trees. Among the world's largest tree-borne fruits, a jackfruit can exceed 120 pounds, almost all of which is edible (the fibrous, stringy matrix is a popular vegetarian meat substitute). Mine were medium-size specimens, weighing about 23 pounds each. When the spiky, leathery skin began to loosen, and a sweet, cloying odor wafted forth, I knew the fruit was ready to eat. I watched a couple of videos on how to prepare the fruit—an operation that requires a large knife liberally slicked with coconut oil lest a latex-like sap ruin the knife with its superglue-like properties. It took me an hour to extract the golden pockets of meat nestled amid the fibrous tissue, each housing a shiny, olive-size, edible seed. The delectable harvest lasted for days and was well worth the effort.

In stark contrast to the fruit, jackfruit flowers are tiny, modest things, and until recently the mechanism of their pollination remained a mystery. Thousands of male and female flowers grow on separate, compound fruiting masses, or *syncarps*, on a single tree. For a syncarp to swell into an edible fruit, pollen must be transferred from the male flowers to the female flowers.

A few earlier studies had sought to elucidate jackfruit pollination, but results were unclear and conflicting. In addition to wind, several flies were considered likely candidates. Thanks to the careful sleuthing of seven researchers from six American institutions, it now appears that pollination results from a three-way mutualism involving the flower, a fungus, and a fly—a tiny new species of gall midge. The fungus forms a weblike film over the male syncarps and is an attractive food source to the fly, both larvae and adults.

Based on a series of investigations—including collecting insects from jackfruit flowers; excluding potential pollinators by placing nets of different mesh sizes around syncarps; measuring the sensitivity of isolated fly antennae to three major scents emitted by jackfruit flowers; and testing the flies' preference for two arms of a Y-shaped tube, one arm of which was connected to a flowering jackfruit syncarp, and the other to the open air—the scientists identified a newly named midge, *Clinidiplosis ultracrepidata*, as the main player in jackfruit pollination. If you happen to be unversed in the *Clinidiplosis* genus, I'm happy to tell you that it contains 104 known species (so far). In English, *ultracrepidate* refers to going beyond one's province; this new species is a long-distance hitchhiker on jackfruit. The study was conducted in Florida, and the authors express admiration that the tripartite fruit/fungus/fly mutualism has managed to migrate intact from the jackfruit's native Asian regions to such a distant cultivated setting.

Hand in Glove

Like tiny midges pollinating hard-to-access jackfruit or cacao flowers, insects and the flowers they pollinate are often finely tuned to each other, and there is a very good reason for this. "Flowers want their visitors to keep visiting other flowers of their own species," Mark Deyrup told me. "This is one reason why the physical appearance of flowers is highly consistent within species (barring manipulation by gardeners) and distinctive across species. Insects who visited different flowers on single foraging bouts wouldn't be spreading much pollen to the right places. Flowers teach generalist insects to be active specialists."

The phenomenon by which pollinators tend to visit flowers of the same species, even where more profitable nectar sources are available, is called *flower constancy*. When pollinators stray, they incur no penalty, but that's not the case for the flower, which can't make use of pollen from another species. In the evolutionary quest for pollinator fidelity, plants tend toward specialized features that favor insects uniquely built to service them. More specifically, because it is nectar that the pollinators are usually after, the plant drives insect-pollinator coevolution by incrementally offering routes to nectar that only one insect can navigate. As we will see, plants have evolved a variety of tactics to compel flies and other insects toward brand loyalty.

The basic approach is to mimic the smells and the visual, behavioral, and other sensory cues that insects use to find food or mates. Then, making flowers hard to reach for all but one or a few specialized insects promotes pollinator fidelity.

When Charles Darwin described flowers whose nectar reservoirs could be reached only through a narrow two-inch flower tube, he famously predicted that there must be some undiscovered insect with an equally long tongue. It was many years after Darwin's death before such insects were discovered. The meganosed fly (*Moegistorhynchus longirostris*) of southern Africa, outdoing its literary counterpart Pinocchio, has a noselike proboscis that is actually the longest mouthpart—a tonguelike drinking tube—of any known fly (see the image in the photo insert). This outlandish appendage protrudes as much as four inches from the fly's face—five times the length of its bee-size body. When folded backward during flight, about half of it trails behind the rest of the insect.

Like other long-nosed flies, the meganosed fly is the sole pollinator to a group, or *guild*, of unrelated plant species that includes geraniums, irises, orchids, and violets. More than 120 known flower species have coevolved longer tubes with flies' longer tongues, allowing the flies privileged access to pools of nectar. In return, these flowers benefit from a near-exclusive pollen transport service, one that minimizes the risk of delivery to the wrong address.

Each plant species arranges its anthers, the male reproductive structures, in a characteristic position. That way, the pollen from each species sticks to the pollinator's body in a distinct but consistent, plant-specific location. The fly becomes an even more efficient courier, carrying pollen from, let's say, three plant species simultaneously, one each on its head, legs, and thorax.

There are risks to such specialization. If long-tubed flowers started disappearing, flies with long tongues could still get nectar from other flowers with shorter tubes, but the flip side doesn't hold true. If long-tongued flies waned, other insects with shorter tongues wouldn't be effective. In parts of southern Africa, loss of wetland breeding habitats is causing declines in long-nosed flies, which in turn is causing flowers in the long-nosed fly guild to produce no seeds, because their pollinator is locally extinct.

Duped and Manipulated

The flower-pollinator mutualism isn't always mutually beneficent. Nature, a great innovator, is always on the lookout for shortcuts. Flies are not routinely beneficial to flowers and pollination systems, and there are flower flies and soldier flies that practice floral larceny by stealing nectar from heliconia flowers without aiding pollination. Pollinating hummingbirds are less likely to visit these flowers when they are infested with the larvae of the nectar-pilfering flies.

More often, it's the plants manipulating the insects. So effective are rewardless flowers in deceiving insect pollinators that these traits have evolved in almost all the major groups of flowering plants, including about a third of all orchids, as many as 10,000 orchid species.

You may have heard of orchids that display plump flowers resembling the abdomens of female bees, which are irresistible to male bees. Drawn in by this plant pornography, the bee makes frantic efforts to mate with the flower, sometimes leaving a sample of his sperm. In the process, he gets stamped (sometimes violently, as we'll soon see) with a packet of the flower's pollen. You might wonder why natural selection would abide males who waste their sperm, but sperm are relatively cheap to produce, and their wastage is not especially rare in nature.

Some orchids direct their seductive tactics at flies. One South American orchid produces flowers with a remarkable resemblance to a female tachinid fly. Among the visual tricks is the flower's stigma, located near the tip of a "false abdomen," which reflects sunlight in the same manner as the genital orifice of the female fly. Male flies pollinate the flowers during their unsuccessful attempts to copulate with the flower. Other flowers are more inclined toward equal opportunity. The dark spots on the petals of a South African shrub attract bee flies of both sexes.

One large orchid subtribe, numbering over 5,100 species, is mostly pollinated by flies. The genus *Trichosalpinx* comprises about 110 species ranging from Mexico and Central America to northern South America. One of the most visible features of these flowers is a dark purple, finely fringed lip, exquisitely calibrated to move only under the weight and momentum of its winged pollinators. However, this structure has another type of motility: it vibrates with the air currents due to the union of the lip base with a thin, flexible ligament.

A research team in Costa Rica found that females of a particular biting midge exclusively visited and pollinated the *Trichosalpinx* flowers. The midges approached the flowers in an irregular zigzag flight and landed on the lateral sepals. They immediately walked to the lip and began to inspect the fringed surface from its tip to its base. Finding just the right spot, they sought and sucked exudates from the flower surface using their fleshy mouthparts. As the tiny midge approached the flower's balance point, her weight triggered a rapid lever movement lifting the lip about 35 degrees upward and slamming the insect against the column. If more than one midge arrived at the trigger point simultaneously, the lever mechanism did not work owing to the excessive weight.

In her ensuing struggle to get free, the midge's back scraped the column's apex, either removing the pollinarium (a pollen-bearing pouch) or, if the fly already carried one, deposited it on the stigma. Mission accomplished, the flower's lip returned to its original position, allowing the midge to fly to another flower or remain on the same one.

It's not a risk-free errand for the midge. In a few instances, the insect was unable to release the pollinarium and get free; thus trapped, she subsequently died in the flower. It's enough to make you wonder whether a seasoned midge might become spooked by the rough treatment and start avoiding these orchids. However, successful pollination happens only if an insect returns to the same kind of flower for another round. Maybe it's fun for them—the fly equivalent of a roller-coaster ride, with drinks on the house.

The jostled midges don't seem to be completely taken in by the flowers. Careful examination of the flowers under scanning electron microscopy revealed no signs that the midges had deposited eggs or larvae. This is probably to both organisms' benefit.

Recall that these midges are blood-feeders, so the orchids are not luring them in with sweet nectar. Rather, they seem to mimic the sensory cues the flies use to access their vertebrate hosts. One clue is that the secretions the midges imbibe from the flower's lip are proteins, not sugars, and it is proteins that the females are after when they go for blood. Another clue is that the flowers attract only females, not males, who don't seek blood meals.

However, a subtler form of mimicry could be at play here. Some of these midges practice *kleptoparasitism*—the stealing of food caught by another predator. They can be found brashly snacking on the prey suspended in a spider's web. Plants that encourage this kind of behavior are called *kleptomyiophiles*, "lovers of thieving flies." The orchids may be attracting the midges by presenting cues that mimic their prey. That fringed lip, which moves due to vibration or wind, might produce a visual effect similar to that of a prey trapped and immobilized in a spider web. These vibrational movements might also aid in dispersing attractive floral fragrances, as has been documented in other orchids. The flowers may initially attract the midges from afar by mimicking the aroma of an invertebrate host. Once the midge comes in close, short-distance cues kick in, tactile (hairy surfaces of the lip), visual (purple color), and mechanical (movement of the lip), that might resemble the body surface of caterpillars or possibly spiders. The small quantities of proteins produced by these flowers suggest that they are "food deceptive." The meager proteins serve as a signal, a tease, to lure female biting midges into the flower and guide them to the pollination point.

Coevolutionary pacts are not limited to just two species. Myrtle flies, for instance, have coevolved highly specific relationships with nematode worms and host plants in the myrtle family, associations that have both parasitic and mutualistic characteristics. Each of some 30 fly species plays host to a single nematode of a species whose entire life cycle revolves around the flies. Mated nematodes infiltrate female fly grubs feeding inside the myrtle, releasing their eggs into the grub's hemolymph. Once ripened, the eggs hatch into juvenile nematodes that are relayed to the host plant when the adult fly lays eggs in myrtle buds or stems. So far, it sounds like the worms are parasitizing the flies. But there's another twist. Once inside the myrtle plant, the worms induce the plant to form a gall—which acts as both a protective chamber and a source of food for the next hatched batch of fly grubs—resources that also benefit the nematodes. So for the price of harboring tiny nematode worms that seem not to threaten their host's lives, the flies get a luxury hotel suite with room service. And they aren't in a hurry to check out: while adults may only live a few hours, the larvae of some gall midges dwell in their comfy quarters up to three years.

Fetid Fly Baits

The bright colors and sweet fragrances of flowers have long inspired poets the likes of Shakespeare and Goethe. Of course, flowers do not exist for our benefit, though florists might argue the point. In the purely biological context, flowering evolved to facilitate the transference of gametes between individuals to achieve cross-fertilization. Less formally, flowers represent mechanisms evolved to exploit third parties, mostly insects, to aid plant sex.

A further rebuke to anthropocentric flower theories is that some flowers are repugnant to our senses. The diversity and abundance of insects, especially flies, whose livelihoods depend on dead and decaying animals or their excrement, presents a way for flowering plants to manipulate insects. In plants that mimic decaying flesh, feces, and decaying fungus, the phony reward is a place to feed or deposit eggs or maggots. It's a costly deception for the flies, because the plant isn't actually providing a suitable site for fly reproduction. These plant mimics combine a suite of hydrocarbons, oxygenated fermentation compounds, and nitrogen- or sulfur-bearing volatiles to produce a unique brew of stinky odors. The deception must be convincing and alluring enough to dupe the insect at least twice, for pollen must be picked up on one flower then deposited on another. These flowers are thought mainly to target egg-laden females, but males are also useful pollinators, for they will visit these flowers while prowling for a swollen mate. Males may get what they're looking for, but the females' offspring will die from malnutrition on the flower. Even the female deception isn't solely exploitative, for many of these flowers provide some form of nutritious reward.

Over eons, the capacity to mimic the appearance and especially the aroma of foul-smelling rotting matter has spawned a botanical cottage industry. Unlike the usual flower-insect pollination systems in which flowers reward pollen-bearing insects with nectar, plants that mimic carrion—like those that offer the false promise of sex—typically provide no compensation beyond sensory allurement. Yet so irresistible are the sights, smells, feels (a warty surface with cilia might suit), and in some cases even the raised temperatures proffered by the deceptive plants that many flies are induced to lay eggs, whose hatchlings will find no suitable food to survive on. Just imagine, if you were a hungry bluebottle fly, stuffed with ripe eggs, how could you be expected to ignore the seductive lure of tainted beef wafting from a nearby three-foot-wide *Rafflesia* bloom?

So refined is the plant's use of "infochemicals" to mimic rotting flesh that their odors reflect not just the olfactory preferences of insects for various types of decaying organic matter, but even to different stages of the decomposition process. Some species of *Rafflesia* preferentially lure females, whose desire to deposit eggs makes them more promising pollinators than males. A research team from Malaysia and South Africa studying five species of these rare plants—which can weigh 7 kilograms (over 15 pounds) and don't bother with stems or leaves—observed only female blowfly visitors at the flowers. When they tested the attraction of the most important isolated compounds from the odors of *Rafflesia cantelyi* flowers to equal numbers of male and female flies placed in a flight tunnel, positive responses from females outstripped those of males by four to one.

These stinky mimics also use some visual tricks to enhance their deception. Flowers that mimic carrion or feces are typically dark maroon, deep red, or dirty yellow, often with contrasting patterns of darker markings on a pale background. Deep reds imitate flesh, and spots or lines may resemble open wounds. Odor and visual cues work in combination. Experiments using model flowers of a South African species, with odor supplied by real flowers, showed that scented black flowers attracted significantly more flies than similarly scented flowers artificially made to look yellow. When scientists added fetid odors to flowers normally visited only by nocturnal hawk moths, these flowers were soon visited by *saprophilous* ("decay-loving") flies, so for some flowers at least, an alluring scent is all that's needed to attract a fly. The evolutionary shift to saprophilous fly pollination may have begun with the production of fetid smells, followed by visual adaptations (color, pattern, and shape) to optimize pollen export and placement between flowers of the same species.

A side note to fly-inspired carrion mimicry in plants is the prevalence of flower gigantism in this assemblage. The single flower of the titan arum, *Amorphophallus titanum* ("giant shapeless penis"), of the Sumatran rain forests can grow over ten feet high, and the well-studied Mediterranean dead-horse arum, which mimics the anal area of a dead ungulate, grows to a foot or more. Among the credible theories proposed to explain the large-flower phenomenon are that such blooms, whose populations tend to be small and scattered far apart in the forest, need to be able to attract pollinators from long distances, and the long-range scent-sensitive capabilities of blowflies fit the bill.

There is a related fly association that is not technically plant-based: flies' relations with fungi. The diverse and successful fungal kingdom has not avoided the notice of flies, many of which feed, take shelter, and reproduce on fungi. The aptly named fungus gnats are a large group, numbering over 5,000 described species, with many more awaiting discovery. As with flies and flowering plants, the relationship is often mutually beneficial. Flies that come into contact with fungi can be effective disseminators of their spores. Some fungi have become so codependent on their partner gnats that, like certain fruits and their pollinators, their spores can germinate only after passing through a gnat maggot's gut.

The relations among flies, flowers, and fungi epitomize a fundamental facet of life on Earth: interdependence. These organisms have been coevolving in their shared spaces for millions of years. Often their associations have become intimately linked, like a lock and key.

The same goes for organisms engaged in sexual reproduction. With rare exceptions, procreation requires two individuals of the same species to pair up and unite their genetic complement. Flies have come up with some creative ways of making more flies.

Chapter 8

Lovers

> Flies love having sex.
>
> —ERICA McALISTER, BRITISH FLY EXPERT

Fly sex comes in fifty shades of brown. Just as the ineluctable need for food has fueled plant allurements for flies, so too has the indispensable need to reproduce spawned a wealth of fly extravagance. As I researched this book, I noticed that many of the most interesting aspects of flies' lives revolve around their reproductive habits. Consequently, it was hard to decide what to include in this chapter and what to consign to a hefty discard pile.

For a great many flies—and many other insects—the adult stage is brief and almost entirely dedicated to reproducing. Lots of adult insects don't even bother eating, instead fueling egg or sperm production with the prodigious amounts of food they ingested during a much longer larval stage. Of those that do feed, many, like the mother mosquitoes who take blood, do so to nourish their developing eggs, not themselves. Growing up, I learned that mayflies were named Ephemeroptera for their fleeting adulthoods, often lasting just a day devoted to mating. An even more extreme example can be found in the abbreviated adult stage of the delicate mountain midges. The time from their aquatic emergence through mating and egg-laying to death is less than two hours. It is a measure of the importance Mother Nature attaches to procreation that so much evolutionary complexity can be invested in an adult fly—all those limbs, senses, and body systems humming away—whose sole goal is to couple with a mate.

How Flies Woo

When Shakespeare's thwarted Romeo laments that "more courtship lives in carrion flies than Romeo," I find myself thinking that the Bard must have had an eye for entomology, for indeed the wooing skills of many flies would provide stiff competition for many guys. True to their diversity, fly courtship takes myriad forms. Robber fly males hover and display over perching lovers, flashing ornaments such as the long fringes on their hindlegs, or their silvery genitalia. Courtship in some long-legged flies includes vigorous waving of their wings and boldly colored legs, and sometimes spectacular backflips and other gymnastic displays. If you're wondering what could drive a fly to do a gymnastic floor routine for a female, chalk it up to choosy females. Male displays have been wrought by generations of discerning dames who favor athletic suitors with elaborate courtship displays.

Signal flies and picture-winged flies combine visual, tactile, gustatory, and probably olfactory and acoustic elements in their multisensory courtships. These flies are named for their prominently patterned wings, each of which they vigorously twist and flex, each independently of the other, in a motion that dipterists call paddling. If this semaphore display clicks and the chemistry feels right, a mated pair may engage in prolonged kissing, locking their lips and exchanging saliva. Males sometimes transfer this fluid to the female's back and imbibe it there; others ingest anal fluids produced by the female. Some male stilt-legged flies are probably spiking their array of allurements with perfumes when they inflate an eversible pouch on each side of the abdomen. Male mosquitoes, which may mate up to eight times, spike their semen with a pheromone called matrone, which acts to suppress the female from seeking further matings.

Be assured, however, male flies are not running the show. Male dance flies might have benefited from perfumes, but they have instead opted for gift giving as a courtship strategy. And with good reason. Adult female dance flies have voracious appetites—the better to fuel egg production—and their mates are on the menu. Males embellish their prenuptial aerial dances with a gift—a freshly captured insect, which can be as big as the male himself. A female will not mate with a male who arrives with no gift. In some species, males simply transfer the prey as is, for their mates to feed on while they copulate on the wing. Others wrap the captured prey in silk released through hollow hairs near the swollen base of their forelegs. This ploy buys the males precious mating time while she unwraps her present. But males are not wholly valiant in their love offerings. Some eat most of their prey first, leaving only a fragment inside the silk balloon. One or two species have taken the bold step of providing the wrapping only, with no prey inside. Others try to get away with provisioning their lovers with a fluffy seed or some other inedible object. Can a female fly feel disgruntled? Females of many species do not themselves hunt, feeding instead on nectar or pollen, so the nuptial gift is thought to provide essential protein required for the development of her brood.

With such lavish gifts on offer, and the possibility of a male meal to boot, competition for mates can cut both ways, with females vying for males. In their efforts to lure a suitable male, some female dance flies deploy a cosmetic trick: they inflate their abdomens to look like they are distended with mature eggs—a fly bustle and an irresistible sight for the aroused males.

Mate attraction is not the only function of courtship displays. Because many species look nearly identical, species-specific courtship maneuvers can also aid species identification. Because mating with the wrong species produces no offspring for either participant, tools for species ID are strongly favored by evolution. (Consider that there are 647 species in the midge genus *Chironomus*; then try to imagine that many species of human, genus *Homo*.)

The mating dances of fruit flies—those familiar little household visitors that gravitate to the fruit bowl—illustrate the point. Each species has an elaborate, multistep sequence. Moves include wing tilts, twitching one or both wings forward, wings outstretched to the side, and rapid wingbeats that create hums of varying tones. Each step must be performed in the right order and at a satisfactory level of proficiency, or else the entire dance is aborted and the flirting reverts to square one (or less). During this dance phase of courtship, one or both flies may take frequent breaks from the action to rest or groom a leg, wing, or antenna; acting nonchalant may be a good dating strategy at this early stage, because the stakes are high when this could be the only sexual encounter of your life. If the dance moves are doing the trick, the action intensifies, eventually leading to physical contact. The male taps and strokes the female's body with his legs and feet. All going well, the male begins kissing his beloved, his soft mouthparts rhythmically pressing against her back and abdomen. Sometimes the petting gets heavier, and body kissing escalates to genital licking. Copulation lasts about two hours. Sex is usually followed by the couple resting and grooming, and occasionally watching television.

A courting fruit fly's wing-generated tones are not merely a by-product of visual display. In fact, they may be the main purpose of the male's wing movements. These "songs" take two forms—a whine like that of a mosquito and a pulse more like a purring cat—and are produced by extending and flapping a single wing. The resulting sounds are exceedingly soft; we must amplify them a million times to hear them. Princeton University researchers discovered in 2016 that male fruit flies adjust the intensity of their courtship songs according to their distance from a female. Crooning louder when your lover is farther away is like our knowing we have to shout when someone is farther down the street. It's a useful ability for a male fly, allowing him to minimize the high energetic cost of singing for females, whose choosiness often requires him to sing for periods that are quite long for a fly. Song-volume adjustment to compensate for distance to receiver is known elsewhere only in humans and songbirds.

I'm sure I have lots of company in having wondered why on earth mosquitoes produce an audible whine. Surely a female mosquito ought to remain silent when approaching a wary creature with a large fly-swatting tail or a lethal pair of hands. Aristotle described the sounds of flies as being "like the opposite of meaningful speech," and concluded, mistakenly, that flies "have no voice and no language."

Closer scrutiny gives meaning to the madness. It turns out that the mosquito's whine—generated by wings beating up to 600 times per second—is not there to annoy its hosts. One clue to its actual function is that in many species only females produce the sound. A further clue is that males receive the sound via their hairy antennae, which vibrate in unison on receiving just the right tone of his species. Sensory cells at the hairs' base transform the vibrations into nerve impulses sent to the brain, which then stimulate the male to fly in search of his mate.

As an extra rebuke to Aristotle, mosquitoes—like fruit

flies—exercise control over their wingbeats. Even though male mosquitoes tend to produce higher whines with faster wingbeats, both sexes actively modulate their flight tones to match their upper harmonics during mating flights.

Mosquitoes' acoustic fidelity is not immune to errors. In some species, the wing tone of immature males resembles that of mature females, which results in some awkward pairings. One entomologist told me that mosquitoes can be drawn to a tuning fork, and in one well-known case the machinery of a power station became gummed up by countless mosquitoes, all males, attracted to the particular high pitch the machinery produced.

A newly emerged mosquito male is, so to speak, prepubescent. His genital equipment requires a day to mature, a process that includes a rotation of 180 degrees. Conveniently, prepubescent males are not only impotent; they are deaf. To be able to hear his girl's love song, a boy mosquito's antennal hairs must be erect, and it is not until genital rotation is complete that hair erection commences. This schedule avoids the frustrating and unproductive scenario of sexually incompetent males uniting with ready females.*

Like their mosquito cousins, frog-biting midges also use courtship songs, and the males also have *plumose* (feathery) antennae to aid listening. A research team from Sri Lanka, the United States, and Panama found that the midges' rapid wingbeats produce tones and harmonics that differ between the sexes. Upon hooking up

*On the flip side, there is at least one case in which sex happens scandalously early for the females, who are effectively raped by adult males when they are about to emerge from their pupae. In the New Zealand mosquito *Opifex fuscus*, males patrol the water surface, rushing to the spot when a pupa surfaces to emerge. He seizes the pupa, causing it to split open. If it's a male, he lets go; if female, he copulates with her. The hapless female becomes pregnant in her first moment of adult life, having had no say in the matter.

into male-female pairs, the two flies adjust their wingbeats to match each other's pitch. Perhaps as a means to repel a wayward suitor, males who were mounted by another male adjusted their pitch away from that of the suitor—a mosquito's way of saying "Get off my back!" The scientists also believe that the courtship function of sound production and hearing may have paved the way for these flies to track frogs by eavesdropping on their calls.

Courtship singing may be the source of a remarkable physical feature in a recently discovered long-legged fly species from the southwestern United States. In all 28 specimens examined, every male *Erebomyia exalloptera* (no common name has been assigned yet) had a left wing that was 6 percent larger than his right, each wing also having a different shape near the tip. Wing asymmetry is unknown in any other winged animal. After all, what could possibly take precedence over being able to fly straight? Sex, that's what. These frail flies measuring about 4 millimeters (0.2 inch) court and mate in dark cavities beneath rocky overhangs along Arizona canyon creeks. Courtship includes wing fanning, which generates distinctive sounds, so an acoustic product of asymmetrical wings seems likely. A hovering male will approach and tap against any dark spot on the rock, and if it's a female, he lands behind her and approaches to within about an inch. Then he extends both wings horizontally and fans them in a series of short bursts. If she doesn't move away, the encouraged male continues his wing fanning while positioning himself above her abdomen and attempting to copulate.

Why would evolution favor such a physical disadvantage as asymmetrical wings? One would think that the penalty of impaired flight would outweigh any possible courtship advantage. But a theory published in 1975 by a brilliant Israeli biologist named Amotz Zahavi offers a potential solution to this riddle. According to Zahavi's "handicap principle," females will choose

The outlandishly long eye-stalks of this male picture-winged fly from Ecuador are a result of generations of fierce competition among males for the attention of choosy females for whom size matters.

(© ROB KNELL)

a mate with the greatest handicap because, paradoxically, this is an indication that he is genetically superior to unburdened males. In other words, so the thinking goes, a male who has survived the odds and reached reproductive readiness in spite of a handicap must be very accomplished at the survival game, and those are genes worth having.

Dealing with Rivals

Discerning female flies are not the only obstacle a sex-hungry male fly must face if he is to breed. Other males are competing for the same prize. Competition for mating rights, typically between males, is widespread in animals, and it can be intense among flies.

Male *Drosophila* fruit flies will spar for up to five hours, and so closely studied are these little flies that scientists have assigned so-called behavioral modules to their competitive maneuvers. They read like something out of a mixed martial arts manual. Here, in roughly escalating order, are the main ones: approaching, wherein one fly lowers his body and moves toward the other; wing threat, wherein one fly quickly raises his wings toward his opponent; lunging, wherein a fly throws himself on his rival; boxing, wherein opponents rise up on their hindlegs and hit each other with their forelegs; tussling, wherein both flies tumble over each other; fencing or kicking; and chasing and holding—as documented by Belgian researcher Liesbeth Zwarts and colleagues in 2012. Like physical aggression in vertebrate animals—but unlike organized bouts between humans—bluffing and ritualized displays are usually adequate to resolve a dispute, so more-violent behaviors that risk injury, such as boxing and tussling, are relatively rare.

Violent aggression can also be avoided when opponents know their place in a dominance hierarchy. Such hierarchies require the ability to remember a previous opponent, and this has been documented in *Drosophila* by a research team of pathologists from Harvard Medical School. They placed pairs of male flies in situations where they fought each other. Then, after 30 minutes of separation, males were re-paired with familiar or unfamiliar opponents. In familiar pairings, there was less fighting than in unfamiliar pairings, supposedly because each fly knew his place. Former losers fought differently against familiar winners than against unfamiliar winners, but they never won against either opponent, nor with naive flies (males with no fighting experience). Winner/winner, loser/loser, and naive/naive pairings revealed that losers dampened their intensity in later fights and were unlikely to advance in the hierarchy. It appears that male fruit flies are learning and remembering their social standing among other males based on their fight rankings.

It should be mentioned that not all aggression between males is over females; food is another cause. Nor does aggression occur only between males. Females will fight over food, especially if it includes yeast, a valuable commodity for developing larvae.

Contests between males vying for females can be remarkably intricate for such small creatures. To illustrate, consider *Prochyliza xanthostoma*, a bone-skipper fly from North America with a winsomely iridescent abdomen beneath perfectly oval wings. They stage territorial combats on or near the newly exposed carcasses of winter-killed animals during early spring snowmelts. Standing tall on their hindlegs, the opponents start by spreading their forelegs and holding each other's front feet, which may help them to assess relative body size—a common predictor of contest outcomes. For up to two minutes, they grapple and box with their front feet and antennae, and butt their greatly elongated heads together—like rutting deer, only at lightning speed. These flies have garnered the name waltzing flies for the coordinated side-to-side movements made by courting pairs. Courtship culminates with the male performing a maneuver befitting an X-rated Olympic gymnastics event. With the wedded pair facing each other, forelegs touching, in an instant the male somersaults over the female's body while rotating 180 degrees, attempting to land on her back. If he scores, he will immediately initiate genital lock, and copulation will last about six minutes.

Male dung flies are less discriminating about whom they target for sex. They have been seen leaping onto the backs not only of other males, but flies of different species, and even gray specks of decay on the lily pads they use for their orgies. Heterosexual mounts, when successful, lead to an elaborate courtship ritual in which the couple rock back and forth for up to 15 minutes. An unreceptive female soon puts a stop to this, but in homosexual mounts, the male who pounced does not quickly dismount on realizing his mistake. Instead, he typically strives to ride his reluctant partner, whose vigorous efforts to displace him may resemble a bucking bronco. Ken Preston-Mafham, who has studied these flies in Warwickshire, UK, believes the upper males stay put not because they are deluded in thinking they might succeed with the fly beneath them, but because they are in a better position to pounce onto the next female who alights on the lily. Male-mounting could also be a way to hone one's female-mounting skills. Male fruit flies that get it on with other males have better chances of mating with females.

Combats and dominance mounts are just two of many strategies male flies use to reduce competition with other males. Dung flies guard their chosen mates, chasing off other suitors. Lovebugs copulate for long periods, making females unavailable for mounting. Vinegar flies transfer a chemical substance that makes the female unreceptive to other mates.

Even the most ardent suitor faces possible rejection by a fastidious female. Female fruit flies have a tactic that leaves little room for a courting male to doubt that he is being spurned: she kicks him in the head with a hindleg. Or she may take a slightly less blunt approach: making her ovipositor (which doubles as a vagina) inaccessible to entry by telescoping it outward, as she would do when depositing eggs. Several volatile hydrocarbons waft from the extruded organ. While a receptive female's perfume can function as an aphrodisiac, in this case her aroma spells rejection—a lady fly's way of using pheromones to tell an amorous male to bugger off.

Rebuffed males not only get the message and back off; they often lose their lust for love. Placed with a receptive virgin female, the rejected male commonly takes no interest. This decline in libido can last hours or days. The phenomenon was first noted by scientists decades ago. Reluctant to attribute emotions like dejection or discouragement to a fly, the term of choice among fly researchers is "courtship conditioning."

Fly Sex

All that courtship and competition are lead-ins for the ultimate payoff: sex. If a male fly's attempts to woo mates fail, or if he gets beaten out by his competitors, then his genes aren't going anywhere. He's a reproductive dead end. The practical efficiency of natural selection ensures that whatever shortcomings a subpar male's genes might have carried, they are less likely to appear in the next generation.

First, let's get the essential anatomy down: the *aedeagus* is an intromittent organ, essentially a penis; and the *bursa copulatrix* is insectese for vagina. The vagina parallel is not only linguistic. A 2010 description of a fly's vagina by three experts reads eerily like that of a mammal's: "The vagina is an elongate muscular tube lined internally with thin cuticle. . . . When the vagina is empty, its wall forms numerous folds. This allows the vagina to become greatly extended [during and after copulation], when containing a spermatophore (a gelatinous mass containing sperm) or, in the case of live-bearing species, a developing egg or larva."

Insect genitals come in a remarkable diversity of shapes, sizes, and structures. There are hairs that tickle, claspers that grasp, inflatable structures, and interlocking devices. Some caddisfly penises come with a *titillator*, a hooklike protuberance believed to stimulate females into mating. One picture-winged fly has a coiled penis that when fully extended equals the length of his body.

Whole books have been written on fly mating parts. There is

an entire book devoted to the genitalia of bee flies of the family Bombyliidae. During an academic-library visit, I leafed through a 178-page treatise for robber fly aficionados titled *On the Genitalia of Asilidae.*

This fixation on flies' private parts is not so much voyeuristic as systematic: their distinctive complexity is often the most reliable feature by which to distinguish otherwise closely related species. This is important to both flies and the scientists who study them. With vast diversity in which closely related species can superficially resemble identical twins, a characteristic knob on the aedeagus or a unique fold in the bursa copulatrix may provide a critical clue to species membership. For the flies, it is no trivial matter that an amorous male fly knows he is about to woo a member of his own species. Time and resources ought not be wasted on wayward liaisons with no actual breeding prospects.

It is probably for this reason that flies have evolved complex genitalia, visual adornments, and structures for gaining and maintaining attachment to one's lover. Take, for instance, the 375-odd species of Sepsinae, a subfamily of antlike scavenger flies. Males often sport conspicuous abdominal brushes and foreleg modifications, some of which deliver tactile or visual stimulation. When Nalini Puniamoorthy and her colleagues at the National University of Singapore bred 27 species of these flies in laboratory cultures, their video recordings revealed that species separation is aided by each type having its own mating style. For example, males may use their hindlegs or midlegs to rub or tap various parts of the female, and their proboscises to "kiss" the female on the top of her head. Many of these mysterious behaviors are of a noncontact variety, as when males curl their tarsi (the equivalent of our toes), or females repeatedly lift their forelegs above their heads.

Successful courtships lead to successful unions, and it is here that fly genitalia really shine. For some flies, a 180-degree genital rotation maneuver forces one partner to lie on his/her back for the genitalia to remain engaged. This explains why one may encounter—as I did on a hike in Maryland—the strange sight of a female fly hanging on her back like a pendulum, legs pointing outward, attached only by the genitals to her mate, who was in this instance perched on a signpost. Don't try this at home.

It says something for the size, strength, and vigor of fly genitalia that one partner can dangle, attached only by the sex organs. The preferred positions for most flies seem to be doggy style, the male mounting from behind while both partners face forward, and end-to-end, the partners facing in opposite directions (see the image in the photo insert), either the same way up or in the 180-degree configuration. If there were a *Kama Sutra* for flies, it would not include the missionary position, which would be impractical for winged creatures who may need to escape danger at any moment. There are, however, awkward situations where the efforts of an unwilling female to shake off a male result in both parties briefly lying on their backs, feet in the air.

How long do flies remain in coitus? Lovebugs (colloquially

For some flies, successful genital interlocking requires that one partner—here a female horsefly in Maryland—rotate 180 degrees, which results in some awkward positions.

(PHOTO BY THE AUTHOR)

known as honeymoon flies) are midges well known in the southern United States, who have earned their name by holding the record for continuous copulation by a fly: 56 hours. (The standing record for continuous sex by an insect is held by a stick insect, whose marathon unions extend to 79 days, a span many times longer than the entire adult life span of many insects, and, I suspect, longer than the cumulative output of most humans.)

All that lovemaking seems to be paying off. Since migrating northward from Central America in the 1940s, lovebugs have been expanding their range eastward by about 20 miles per year. They reached Pensacola by 1949, Tallahassee by 1957, Gainesville by 1966, and South Florida by 1975. These amorous flies' association with car windshields—against which they may get splattered to a degree that can make safe driving difficult or impossible—is not a chance thing; they are attracted to the ultraviolet light emanating from highways, and to a mystery ingredient in car exhaust fumes. Copulating pairs have been found as high as 1,500 feet.

When a male lovebug mounts a willing female—often after he has won a wrestling match with up to ten other male suitors—he does so in the usual fashion, by climbing on her back and uniting his complex genitals with her complex genitals. Three muscular, interlocking claspers and valves are deployed, resulting in a firm attachment. Once docking is complete, he performs a 180-degree rotation, resulting in a head at each end. In this position they take flight. Both spouses flap their wings, which led scientists at the University of Florida and the USDA to ponder whether the male's efforts aided or impeded forward flight. They devised a simple yet clever method to resolve this: by comparing average flight speed of single females with that of couples. The speeds were 44 and 51 meters per minute (1.5 and 2 miles per hour), respectively. It appears that either female lovebugs are supercharged by sex, or coital males can fly backward.

Other scientists have determined that a male lovebug has transferred all of his sperm to his mate within about 12 hours. Why, then, does he remain attached for a day or more? The most likely explanation lies with a phenomenon called sperm precedence, wherein the last male to mate fertilizes most of the eggs. Prolonged copulation is an anti-cuckolding strategy. By staying inside his mate, a male deploys a very direct method to block entry by others.

Genitalia on the Move

It has not gone unnoticed, especially by a growing cadre of female dipterists, that most of the research on fly genitalia has focused on male flies. "In comparison to male structures, the external and internal female reproductive tract is largely a 'black box' that remains poorly known," write two leaders of research into female fly genitalia, Nalini Puniamoorthy and Marion Kotrba at the Bavarian State Collection of Zoology in Munich. In fairness, external parts (males) are more easily studied than internal parts (females), but careful dissections and microscopes can be combined to allow effective study of inner structure. In a 2010 study, Puniamoorthy and Kotrba—with Rudolph Meier of the National University of Singapore—described rapid evolution of female genitals in 41 species of black scavenger flies.

Females need to rapidly evolve mating structures, if for no other reason than that males are rapidly evolving theirs. After all, what would be the benefit of one gender evolving in one direction if the other gender were not keeping up? Cross-gender changes *have to* co-occur, for natural selection will weed out any males whose modified genitals are not accommodated by female genitals. Hand fits glove.

But females have reproductive options unavailable to males.

For one thing, promiscuity broadens a female fly's options for which sperm to assign to her eggs. As Puniamoorthy and her team note, "Studies indicate that females can influence paternity by differentially storing sperm from various males in separate sperm storage pouches and controlling which pouch is used for fertilizing eggs." That's a pretty impressive skill, but how they choose and what qualities of ejaculate they are discriminating among remain a mystery.

Puniamoorthy and her colleagues identified two features of the female reproductive tract that are evolving quickly. One of these, the *dorsal sclerite*, a hardened region of the vaginal wall containing the openings of ducts that receive and store sperm, interacts with the male phallus. The second, a sperm-storage organ called the *ventral receptacle*, is probably where fertilization happens. "Both structures are potential targets of post-copulatory sexual selection in Sepsidae." That is to say, females probably have control over which male(s) get to fertilize their eggs.

Meanwhile, males appear to be trying to influence female choice with stimulatory tactics. For example, it seems likely that a male's intromittent organs rubbing against the dorsal sclerites serve as courtship signals to influence postcopulatory female choice. Might males be using pleasure to sway females to choose their sperm over that of another male? I haven't met an entomologist comfortable with such an interpretation, but we should not rule it out. Incidentally, both sexes of scavenger flies have large genital glands that release an aroma smelling like a mix of lemon and thyme, leading to one suggestion that they instead be called scented flies.

How can scientists see what's happening on the inside of a pair of fornicating flies? One research team studying five species of tsetse flies used three techniques: (1) they flash-froze mating pairs, then dissected them; (2) they artificially stimulated males;

and (3) they viewed copulating pairs with a new X-ray technique that allows events inside the female to be recorded in real time. The scientists admit that their data "almost certainly give only an incomplete view of this complex, previously hidden world."

During an intimate lunchtime conversation, a friend of mine was asked if he talks during sex. "Only when I'm by myself," he replied. Not so with flies. Enter the *copulatory dialogue*. A recent study of "internal courtship" in tsetse flies found that females are signaling to males during copulation. Daniel Briceño and William Eberhard, in a 2017 paper, describe two apparent female signal types: wing vibration and body shaking. A female vibrates her wings when the male squeezes her abdomen rhythmically with his powerful genitalia. The result is usually shorter squeezes by the male, and this coordination suggests that the wing vibration serves to signal the male to interrupt his squeezes without forcibly dislodging him. Female body shaking was usually elicited by especially strong body squeezes by males.

Males lengthen their script in the copulatory dialogue by using their muscular genitalia to "perform dramatic, stereotyped [distinctive to the species], rhythmic movements deep within the female's reproductive tract and in inward folds of her external surface." The movements appear not to help the male anchor himself more securely to the female. What are they for, then? Titillation, perhaps? The authors conclude that, even by a conservative count, "a female [tsetse fly] may sense stimuli from the male's genitalia at up to 8 sites on her body during some stages of copulation."

Is It Fun?

Might flies, as I've already suggested, be enjoying sex? I encourage you to watch an online video by Rob Curtis showing two signal flies (named for the way they communicate by waving their wings like semaphore flags) mating on a leaf. The male approaches the female from behind—her abdomen temptingly swollen with unfertilized eggs—and carefully mounts her. The female immediately stops grooming herself. Their genitals seem to take on lives of their own, inflating, telescoping, and flexing. Once the genitals are locked in, the male makes repeated abdominal thrusting movements. Both participants stroke their partner's abdomen with their hindlegs. At intervals, the male leans forward over the female and she raises her head, leading to a kiss, during one of which there is an exchange of fluids. If you can get past the fact that these are insects, the action is disconcertingly humanlike.

Are they enjoying it? Why not? After all, sex is top priority for adult flies, who may live only for hours and whose chief if not sole life goal is to procreate. From a purely genetic perspective, it is no less important that fly genes make copies of themselves than that human genes do, so we may expect nature to ensure that flies are highly motivated to mate. What better motivator than pleasure?

A study published in 2012 hints that sex might be rewarding for flies and reveals an aspect of fly behavior strikingly reminiscent of our own. Given their attraction to rotting, fermenting fruit, fruit flies encounter plenty of alcohol in nature, and they have some tolerance to it, but like us, they benefit from limiting Happy Hour to an hour. They tend to become inebriated when their blood alcohol level reaches about 0.2 percent (our common legal driving limit is 0.08 percent).

Male fruit flies were paired with females who were either receptive to males or rejected them because they had already mated. Later, when given the choice between a solution laced with alcohol and a nonalcoholic solution, the sexually frustrated males chose to consume higher levels of alcohol than the mated males. This echoes our own tendency to turn to the booze when we're feeling disappointed.

But to really get at the question of whether flies enjoy sex, it might help to home in on an easily observed aspect of sex that is clearly rewarding (at least in humans). What about male ejaculation?

To isolate ejaculation from other potentially pleasurable elements of sex, a team of researchers led by Galit Shohat-Ophir at Israel's Bar-Ilan University genetically engineered male fruit flies with abdominal neurons that are activated by red light. These neurons produce a chemical, corazonin, that stimulates ejaculation. As a result, about thirty seconds after these males enter a space illuminated by red light, they have a fly orgasm. They proceed to ejaculate about seven times a minute for up to three minutes.*

When the researchers placed normal and genetically engineered flies in an unlit chamber, the insects distributed themselves randomly. But when they placed the flies in a chamber half of which was dark and half illuminated by red light, the GMO flies began to show a strong preference for the red-light district.† Normal flies (and female flies) showed no preference. These results suggest that the modified males enjoyed it there because of the climaxing. It's worth mentioning that the modified flies weren't merely attracted to red light, since fruit flies cannot see red. They see light, but it just isn't red in color.

To bolster the conclusion, a further test was done in which flies were trained to associate one of two different odors with corazonin activation leading to ejaculation. When these flies were then tested in an arena with both odors presented simulta-

*We might consider that this sort of thing might not be wholly pleasurable.
†Credit to science writer Ed Yong for the pun.

neously, they approached the odor that they associated with ejaculation. Control flies who had been exposed to the odors without any corazonin activation showed no preference between the two smells.

The team also found that flies who had ejaculated shunned food laced with alcohol, compared with nonclimaxing control flies. "If the reward system is saturated, ethanol [alcohol] is no longer perceived as a reward," says Shohat-Ophir. This complements the findings of an earlier study that Shohat-Ophir coauthored, which found that sex-deprived flies will more readily turn to alcohol than will males who have successfully mated.

Science journalist Andy Coghlan neatly summarizes the research: "Male fruit flies seem to enjoy ejaculation as much as men do . . . [and] their 'orgasms' seem to be satisfying enough to reduce their craving for other rewards such as alcohol." It also creates positive associations with other stimuli, like smells. As Shohat-Ophir notes, "This sexual reward system is very ancient machinery, conserved from simple organisms all the way to us."

There's another benefit to an active sex life if you're a fruit fly: longevity. A 2015 study found that frustrated male fruit flies exposed to female pheromones with no opportunity to mate became stressed and prone to starvation through loss of their fat stores. These deprived flies lived shorter lives. "It may not be a myth that sexual frustration is a health issue," concluded University of Michigan researcher Scott Pletcher.

As I read about this research, I couldn't help but wonder—as does ethologist Marc Bekoff in a blog post—about the apparent lack of research on sexual reward in *female* fruit flies. Procreation is just as important to them as to males, and even though a female cannot potentially produce nearly as many offspring as a male—because even the smallest eggs are much bulkier than the largest sperm—sex is still indispensable to female flies with regard to individual reproductive success. Do female fruit flies enjoy it? Might they have orgasms? Male orgasm is easier to observe than female orgasm, but as we've seen, female fly sexuality is not beyond our reach.

What we currently know is not especially encouraging. There is some evidence that sexual intercourse has a muting effect on libido in female fruit flies. A 2019 study by American and Canadian researchers found that proteins in sperm and seminal fluid left by an earlier mating lower a female's receptivity while stimulating egg production and egg-laying. Furthermore, the sensory experience of copulation also dampens females' interest in subsequent males—a so-called *copulation effect*. So it seems that both females and males may be influencing the propensity for these flies to mate only once, which happens to benefit the reproductive success of both partners. This certainly doesn't close the door on the possibility that females might also be enjoying sex, but they appear wired to avoid a repeat performance.

And with good reason. Reproductive senescence is a well-documented phenomenon in humans; fertility declines dramatically during a woman's fourth decade. Reproductive senescence also happens to female (and surely also male) fruit flies. It is accompanied by declines in fecundity, fertility, offspring longevity, and female receptivity. If you're chuckling to think that a fly could experience lower libido, don't laugh too hard. The dopamine system, which is linked to pleasure in vertebrates, including us, appears to influence female sexual receptivity in fruit flies.

Whatever you may think about fly reproduction and whether flies enjoy it, you may at least take pleasure in the fact that such things are cutting-edge science these days. It will be interesting to see what new discoveries come to light in the years ahead, now that scientists are open to researching whether sex is fun for a fly. If insects do get a kick out of sex, then there is more pleasure in the world than we had thought.

Let us not forget why flies mate in the first place. It's all about procreation, and there, too, we encounter some surprises. Here's one. True to their diversity, flies use a variety of birthing styles. I remember borrowing a book about snakes from my middle-school library and learning, with some pride, three interesting-sounding words: *oviparity*, *ovoviviparity*, and *viviparity*. In researching the current book, I discovered that these terms also apply to flies. Oviparity is simply the production and deposition of eggs, and it's the method most flies use. Ovoviviparous flies do what oviparous flies do, with the crucial exception that the larvae emerge from their eggs while still inside the mother, usually shortly before deposition. There is some urgency to subsequently letting them out because occasionally the ungrateful little tykes will start eating the mother from the inside. Viviparity involves the birthing of offspring that develop internally from an egg without a shell-like covering. During gestation, the embryos are kept in a uterus-like structure and nourished with microbe-rich fluid delivered from a milk gland. It sounds a lot like mammalian gestation. Viviparous insects produce fewer young than do oviparous ones, probably because the gestational investment in each baby is greater, resulting in a higher chance of survival. A few species, like the tsetse fly, have taken this strategy to its limit, producing in each pregnancy just one larva, which at delivery measures nearly three quarters the length of the mother's own body.

Finally, a word about aesthetics. Flies have different tastes from ours, and we may regard the idea of courting and mating on a moist pat of cow dung or a blob of waterfowl excrement as the height of disgusting. But we might reserve a little empathy, not to mention awe, for the varied ways that different creatures, large and small, get primed for sex. There is no need for judgment over the sexual norms of another species. Indeed, given the bizarre suite of fetishes for which humans are known, we are hardly the gatekeepers of sexual purity. The same human who may grimace at the thought of a fetid fly love nest might not hesitate to get it on in a giant bowl of chocolate pudding.

Part III

Flies and Humans

Chapter 9

Heroes of Heritability

Am not I
A fly like thee?
Or art not thou
A man like me?

—WILLIAM BLAKE, 1794

In the remaining chapters, we will explore the human-fly nexus. By what means is a fly the deadliest enemy of humans, and what are we doing about it? How do flies help us solve murder cases, and why do surgeons sometimes turn to maggots for healing their patients? And how does a fly help us to understand evolution and the building blocks of life? Let's begin with the last of these.

Ask scientists to name the one organism that has contributed more than any other to our understanding of genetics, and most will choose a fly. The fruit fly, *Drosophila melanogaster*—"dew-lover with a black belly," in reference to the male fruit fly's inky hindquarters—is the dipteran darling of genetics research.

Fruit flies (technically they are vinegar flies) are diminutive insects: as you will appreciate in the likely event you've seen them in your kitchen or elsewhere in the home, a dozen could fit comfortably on your thumbnail. Having arrived in the Caribbean via slave ships from Africa and southern Europe, by the 1870s the fruit fly had made its way to New York, Philadelphia, Boston, and other major cities in North America, abetted by a burgeoning post–Civil War trade in rum, sugar, and bananas and other tropical fruits. With plenty to eat and lots of suitable human-made habitat to occupy, the little fly soon established itself in its new realms.

The story of the fruit fly as the champion of animal genetics research began around 1900, when a Harvard graduate student, Charles W. Woodworth, began breeding them for embryological studies. A few years later, a zoology professor named Thomas Hunt Morgan noticed a spontaneous change in eye color in the fruit flies he was breeding at Columbia University, and the fruit fly's scientific career took off. Between 1910 and 1937, the number of fruit fly laboratories in the United States and Europe grew from 5 to 46.

Today, the fruit fly has the distinction of having consumed more printer's ink than any other insect, with the possible exception of the honeybee. In addition to some hundred thousand papers in scholarly journals, hundreds of books and manuals on *Drosophila* genetics have been published. There is a journal dedicated to the study of flies. Appropriately titled *Fly*, it is focused exclusively on *Drosophila* research. If you want to know how developmental temperatures affect the intestinal biome of a fruit fly, or how to accelerate your lab's production of fly genomic DNA using a paint shaker, this is the place to find out.

It is a measure of the scientific tendency toward specificity that *D. melanogaster*, for all its popularity in the laboratory, is just one of nearly 4,000 described species in the megadiverse genus *Drosophila*. Most species in this group feed quietly on decaying plants and fungi. Others occupy more violent niches related to parasitism or predation. Their varied lifestyles include preying on blackfly and midge larvae, devouring dragonfly eggs, slurping slime fluxes, cohabiting with crabs, and feasting on frog embryos still inside their egg masses. With company like that, the frugivorous habits of *D. melanogaster* are decidedly demure.

One of the benefits of focusing on one species is that scientists can more readily build on the discoveries of their colleagues. "Everything in modern genetics, from gene therapy to cloning to the Human Genome Project, is built on the foundations of early-twentieth-century fruit fly research," reports Martin Brookes in his 2001 book *Fly: The Unsung Hero of 20th-Century Science*. "Radiation is harmful to flies, and flies helped us discover that X-rays are bad," fruit fly geneticist Kelly Dyer told me. "Many of our discoveries about inheritance were from fly research, and few people realize that a lot of what we study about cancer is from flies." Among the sweeping fruit fly–assisted revelations that made the 20th-century Russian American geneticist Theodosius Dobzhansky famous were that wild populations contain a reservoir of genetic variation, that genes are the currency of evolutionary change, and that wild populations (of animals with short generation spans, like flies) can evolve in just a few months.

The early 1980s saw the emergence of powerful new tools of genetic manipulation that allowed, for example, the isolation and cloning of individual genes and the decoding of a gene's sequence of DNA letters. Then in 2014, scientists perfected CRISPR-Cas9,* a revolutionary gene-editing technique. CRISPR co-opts the cellular DNA-repair machinery, allowing geneticists to swap in and out any sequence of genes at will, right down to the level of a single base pair. The CRISPR system has generated enormous excitement in the scientific community because it is faster, cheaper,

*Short for: clustered regularly interspaced short palindromic repeats and CRISPR-associated protein 9.

more accurate, and more efficient than prior genome-editing methods. Such is CRISPR's power that its development is widely considered to be a legitimate contender for a Nobel Prize.*

Because many genes, like fossils, are remarkably well preserved over time, CRISPR techniques have broad potential application throughout the pantheon of life. Take, for example, the CREB gene, which is crucial for long-term memory in fruit flies: it is also found in sea slugs, nematode worms, rats, mice, and humans. Disrupt the CREB gene in mice, and the mice have only short-term memories, which are not retained. More surprising than that, splice an extra CREB gene into a fruit fly genome and the fly demonstrates a greatly enhanced memory, learning, for example, to associate a particular odor with an electric shock after just one trial, instead of the usual ten or so.

So far, fruit fly research has earned seven Nobel prizes. Today, fruit flies are being used in research on—among other things—aging, toxicity, immunity, epilepsy, neurodegenerative disorders like Parkinson's disease and Huntington's chorea, microbial diseases like Ebola and cholera, and the evolution of sentience. There are about a hundred thousand strains of *D. melanogaster*, carrying just about any mutation you can imagine. The diversity of mutations wrought on these insects by geneticists is reflected in the animals' creative, often irreverent names: the Methuselah mutant is stress-resistant and tends to live longer, whereas the Drop Dead, Sponge Cake, Swiss Cheese, and Egg Roll mutants carry hereditary diseases resembling patterns of human brain degeneration. The Ken and Barbie mutant lacks external genitalia,

*As this book was going to press, the 2020 Nobel Prize for Chemistry was awarded to Emmanuelle Charpentier at the Max Planck Unit for the Science of Pathogens and Jennifer Doudna at the University of California, Berkeley, for developing CRISPR.

Cheap Date is especially susceptible to alcohol, and the very short-lived Tin Man lacks a heart.

A Visit to a Fly Lab

Eager to see a modern fruit fly genetics lab in operation, I arranged to meet Kelly Dyer in her clean, windowed corner office at the Davison Life Sciences Complex on the University of Georgia (UGA) campus. She welcomed me to a chair beside her desk. A bookshelf containing mostly books about flies stood against one wall. Another bookshelf hinted of a person who doesn't always have time for a sit-down lunch: it held three apples, a jar of peanut butter, and some granola bars. No wonder. Dyer juggles teaching, research, writing papers, attending committee meetings, and running a graduate program for UGA's Genetics Department, which supports over 30 faculty members and some 50 PhD students.

"Where do you see genetic research on *Drosophila* going?" I asked Dyer.

"There are two developments that have made genetic research more powerful: one is that we can make transgenic organisms, so it is much easier to manipulate genomes than it used to be. And two is that we can now sequence genomes quickly and easily. For instance, scientists went out and got 200 wild flies, bred them in the lab and sequenced an array of 200 genomes. The result: a comprehensive resource called the Drosophila Genetic Resource Panel (DGRP). It is really useful for understanding the genetic basis of traits, because you can look at the different genotypes [the organism's genetic blueprints] and cross-reference them with phenotypes [observable expressions of genotypes], like toxin tolerance. Then we can 'knock out' these genes and see what physical manifestations occur."

Dyer led me on a tour of her adjoining lab, a room measuring about 40 by 40 feet, with blacktop workbenches, rows of instruments, drawers of lab tools, and trays of clear plastic vials, each measuring about three inches tall by one inch wide and capped with a white plug of cotton. At the bottom of each vial was a bed of fly food medium—a mixture of molasses, brewer's yeast, commercial fly formula, and water. In the space between the medium and the cotton one could see fruit flies in various stages of development. Some vials contained eggs, others larvae, pupae, or adult flies. Some contained a combination.

Dyer explained some of the other equipment she and her students use in their fly genetics research. Fancy-looking micropipettes lay in a basket. These handheld devices first suck up and then dispense exact measures of liquid by the milliliter. Each has

A plastic vial with food medium on the bottom and a foam cap on top—standard equipment in the fruit fly lab.

a dial to set the desired quantity, which is dispensed when the button at the top is depressed with the thumb. Price tag: $300 each.

On a neighboring bench lay a pile of clear plastic pouches, each containing a rectangular sheet stamped with 96 tubular depressions, in the manner of an ice cube tray. Dyer opened one.

"We dispense a quantity of liquid buffer medium into each tube, then we add a genetic sample."

"How do you obtain a genetic sample?" I asked.

Dyer handed me a solid blue plastic wand about the size of a drinking straw but with a rounded, cone-shaped tip molded to fit perfectly into the well of any one of the 96 tubes.

"After being anesthetized with carbon dioxide, a fly is dropped into the tube and macerated with the wand before being centrifuged to extract the full complement of its DNA. The solution turns red because of the pigment in the eyes." (Insect blood is not red.)

"This is our fancy centrifuge," Dyer said with a playful smile. She was holding an ordinary salad spinner with a manual depression pump on top. With the high cost of lab equipment, scientists are always looking for cheaper ways to do their work.

"To look at flies, we use dissecting microscopes, and we use carbon dioxide as an anesthetic."

Dyer drew my attention to a small chart on the wall. It depicted male/female pairs of about twenty *Drosophila* species. She pointed to a section of the chart.

"These are some of the species we'd find in mushrooms. You can see that they're superficially alike, but there are many variations."

I peered closer and could see that some species had distinctive patterns of spots on their wings, others variable abdominal patterns of color and shape, some fused, some separate.

Turning to the workbench nearby, Dyer picked up a vial, turned it upside down, then plucked out the cotton plug and vigorously tapped the vial about a dozen times onto the surface of a specially designed rectangular stage. The stage, with the dimensions of a dishwashing sponge, was made of a hard, finely porous material through which carbon dioxide was flowing via a tube leading from a nearby metal gas canister. The half-dozen flies in the vial dropped immediately onto the stage, where they twitched for a few seconds before becoming still. Dyer then tipped them onto the viewing stage of the dissecting microscope.

"You can't leave them on the carbon dioxide stage forever because they will die."

"How long is forever?"

"About twenty minutes. In some species, there's actually a virus that causes flies to die immediately when gassed with carbon dioxide."

I had read that carbon dioxide is odorless (unlike the ether used in my undergraduate genetics lab), but I couldn't resist leaning forward for a quick sniff. I felt a tingling in my nostrils, as if my body was sending me a warning.

"I've actually never done that!" said Dyer.

Inevitably, some flies escape the apparatus. On the day of my visit, several were repeatedly swept off the carbon dioxide stage by mysterious gusts that drew sighs of frustration from Dyer. In a few minutes, they would recover, winged fugitives exploring the hallways of the building. With the omnipresent aroma of fly food here, I doubt many stray far.

Working with fruit flies for two decades has done nothing to dampen Dyer's admiration of them. Peering into the dissecting scope to gaze at a newly knocked-out batch, she crooned, "Look at how beautiful these flies are! These flies are just incredible. So pretty!"

I focused my eyes into the scope. There lay seven flies—each a unique, tiny, exquisite jewel of life. They exemplified perfection.

These were wild-type (nonmutant) flies, with prominent pink compound eyes peering out expressionlessly. Spongy mouthparts resembling megaphones gave no hint of the small pair of black, hooked, scraping claw-mouths of the amorphous larvae they had been just a week ago. A perfect row of tiny square spots lined the edge of each fly's belly, and several other spots peppered each wing in a symmetrical array. Flexed legs, twitching slightly, were decorated with neatly arranged spines. The completeness of the adult flies, their superb symmetry, the sheer beauty of their details, reminded me that I was gazing at the product of billions of generations of evolutionary honing.

The next batch Dyer readied were mutants with a genetic aberration on every chromosome. Their eyes were brown, and much smaller than their untampered counterparts.

"All those mutations make them pretty unhealthy," said Dyer.

This was a good time to pose a question I'd been wanting to ask Dyer: "Do you think flies are sentient? Do they have any experience?"

"We respect our study organisms, and we think our flies seem to have a good life in the lab because they have plenty of food and no predators. One day in the lab when I was a graduate student, I noticed a freak mutant fly that didn't have any external genitalia. No anus. Just smooth. I said to Jerry (my supervisor): 'Hey Jerry, this is really cool, come have a look.' Jerry took one look at that fly and he said, 'Oh, you'll have to kill it, it's in excruciating pain.' I said: 'Why do you think that?' and he replied, 'Well, if you couldn't excrete anything from your body, how would you feel?'"

I couldn't help feeling some sympathy for these flies. In a genetics lab, their fates are so utterly under our control. But then, the fate of all flies is profoundly tenuous. In the numbers game, so few make it to adulthood. And in the nourishing realm of a genetics lab, their odds are infinitely greater.

The Morgue

Unlike Kelly Dyer, I was never cut out for genetics. In addition to the field's requirement of mathematical competence, which I generally lack, I do not enjoy confining animals, and short of self-defense I am loathe to kill them. And so it was not entirely coincidental that, 36 years before visiting Dyer's lab, I had conducted a minor act of animal liberation in an undergraduate genetics course.

Over several weeks we cross-bred different strains of fruit flies to observe the distribution of phenotypes in the resulting offspring. The well-studied phenotype we were using was eye color. Specifically, we combined normal red-eyed adult flies with mutant ones bearing white eyes. The larvae and pupae were housed in small plastic vials just like the ones I would see many years later in Kelly Dyer's lab, each containing a yeasty medium. Sponge caps allowed the passage of air but not flies. For several weeks the lab held the dense aroma of unbaked, slightly rancid bread.

A week after the flies had mated, plump little maggots could be seen worming their way through the yeasty sludge. A week later, the vials were spattered with inert pupae, and by week three, each vial had sprung to life with the emergence of adult flies. Some stood, some walked, and others buzzed about their tiny prisons.

In the name of education, these little insects were not destined to enjoy more spacious realms. A gust of ether soon knocked them unconscious. Our instructions were to tip the anesthetized flies onto a piece of white paper, record the number of red- and white-eyed individuals under a dissecting microscope, then dump the still-stunned subjects into a little glass dish of oil sitting on each of the lab desktops. The morgue, as these dishes were called, was to be the flies' final resting place. I noticed that most of the morgues in the room already contained hundreds of flies, presumably dumped there by other students in other undergraduate labs. I still remember their eyes staring balefully from their murky graves.

Unhappy at the prospect of playing Grim Reaper to these spritely little creatures, I hatched a simple rescue plan. Having counted my batch, I nonchalantly tipped them onto the black desktop, against which they were nearly invisible from a distance.

As I continued to jot down data, I kept one eye on the small dipteran heap a few inches to my right. Within a few minutes, I detected some signs of movement: a wing twitch here, a flexing leg there. A minute later, some of the flies were clearly emerging from their stupor. A few spun in circles on their backs, breakdancing as their wing engines sputtered to life. Others managed to get onto their feet and stumbled about like diminutive drunkards. I was mesmerized. Their behavior was so humanlike, and considerably more interesting to me than measuring eye-color ratios according to established genetic knowledge.

I don't know how a fruit fly's experience of ether intoxication might compare with my own. I was anesthetized with ether in a Toronto hospital in 1972 to reset a dislocated ankle after a tobogganing accident. I still remember its distinctive, sharp essence, and the grogginess that accompanied reawakening. Were these fruit flies feeling groggy? We can never know for certain, but as I watched these ones regain their wits, it sure looked like it.

As the flies fully recovered their faculties, some began to lift off. I watched their tiny forms recede and vanish into the cavernous classroom. I don't know if it was the satisfaction of sparing their lives or the exhilaration of committing an act against authority, but as each tiny wisp of life made its maiden launch into the great unknown, a tiny piece of my soul went with it.

Rovers and Sitters

Within a year of that genetics class, in the same building, I took an undergraduate behavior course with Marla Sokolowski, a behavioral geneticist now at the University of Toronto. While researching this book, I arranged a Skype call with Marla. Early in her career—while still an undergraduate—she discovered that fruit fly larvae forage in two distinct, genetically coded styles, a phenomenon formally referred to as the *rover/sitter polymorphism*. Rovers plow restlessly through their food, which in Marla's lab is typically a paste of yeast and water, or overripe fruit. Sitters are more passive, preferring to munch away at food within reach before inching forward.

It is a paradox of knowledge that the more we know, the more unanswered questions we are faced with, and Sokolowski and her students have published over 65 scientific papers on fruit fly behavioral genetics, among dozens of others.

The difference in locomotion between rovers and sitters disappears in the absence of food and is therefore thought to be foraging-related. Marla told me how adversity also influences the foraging behaviors. When faced with early adversity in the form of food limitation, sitters will take more risks. Specifically, they will start darting into the middle of a food-laden petri dish in the manner of a hungry rover. The rover/sitter polymorphism helped to foster a new subdiscipline: behavioral genetics.

That genetically coded behavior is not inflexible has spawned a revolutionary change in our view of the old nature/nurture debate, which had pitted genetic against environmental influences in the expression of personality traits, health trajectories, athleticism—you name it. The reality is that who we are is influenced by both genes and environments, and furthermore that their effects are codependent. Gene–environment interactions are therefore better characterized as nature through nurture, rather than as nature versus nurture.

Like that of many *Drosophila* researchers, some of Sokolowski's work explicitly seeks to advance the human condition. "Mutants that disrupt many of the [flies'] social interactions . . . might provide good candidate genes for . . . autism," she writes in a 2010 paper. "Aggressive interactions [of flies] that repeatedly end in defeat could be used to model chronic defeat syndrome found during depression and identify candidate genes for this disorder." But, she continues, "one needs to be cautious because comparisons between animal models and human social disorders should be based on similar genetic and physiological mechanisms, not just whether the behaviors appear similar." The worry here is that the idea that "complex behavior in mammals derives from simpler modules of behavior in simpler organisms" (e.g., flies) is tenuous.

Fruit Flies on the Move

Relating fruit fly phenomena to human health is not a direct research concern of Patrick O'Grady. I met O'Grady in his tidy, windowed office on Cornell University campus in Ithaca, New York. A well-built family man in his forties who often bikes to work, he was wearing a Hawaiian print short-sleeved shirt, perhaps as a nod to the large chunk of his career he has spent studying flies in Hawaii.

An early introduction to fly genetics enamored O'Grady with fly diversity, which in turn led to travels into Mexico and South and Central America to collect flies for his PhD. He produced a phylogeny for the entire Drosophilidae family, which today numbers about 4,200 described species, over 80 of which were first described by O'Grady.

Hawaii is a hotbed of fruit fly diversity. "There are about a thousand species endemic to Hawaii, a quarter of the family's worldwide diversity," O'Grady told me.

Compare that with a 1945 book on Hawaiian insect fauna, whose authors at the time boldly boasted that there may be as many as 250 species of *Drosophila* in Hawaii.

O'Grady continued. "The Hawaiian *Drosophila* are just spectacular. There are two major groups, one of which is characterized by their sexual dimorphism. Males are really showy. It's like the Irish Elk or Bighorn Sheep scaled down to fly size. Some of these guys have modifications to their mouthparts, others their forelegs, or their wing patterns."

If you Google Hawaiian fruit fly wing patterns, you will find a cornucopia of streaks, stripes, speckles, and blotches. Fashion designers, take note.

"A couple of species butt heads in contests over females, leading to the evolution of broad heads. Almost all of these species form leks" (specific locations in the habitat where, like some birds, males engage in contests for females).

Now, you may be thinking that the high Hawaiian quota of fruit flies is just because we haven't scoured the planet for their cousins, whose diversity will turn out to be far greater than we had thought. Perhaps, but not according to O'Grady.

"One theory for the megadiversity of Hawaiian fruit flies is that these flies are much older than the islands themselves. Hawaii is basically a conveyor belt archipelago with a plume of lava at the Pacific plate that continually moves to the northwest, a few centimeters per year. Eventually, the island falls off the plate and another island pops up."

Later, when I looked at a map of the Hawaiian islands, I could see that they fit O'Grady's description beautifully—extending in an orderly row toward the northwest, becoming generally smaller until they are engulfed by the ocean. The entire chain represents about 60 million years of plate tectonic shift. It's estimated that the first Hawaiian island inhabited by fruit flies is now submerged somewhere near the Midway Islands, over 3,000 kilometers (1,860 miles) to the northwest.

"We think *Drosophila* got to Hawaii about 25 million years ago and they've been jumping along this conveyor belt as they go. And every time a new island forms, the flies colonize it and you have a whole new process of speciation taking place in isolation from the neighboring island."

There are other factors contributing to Hawaii's fruit fly bounty.

"There is also a lot of habitat subdivision on these islands," O'Grady told me. "One island isn't even necessarily made up of a single volcano. The big island of Hawaii, for instance, has five volcanic peaks. Trade winds ensure that one side of these peaks gets much more rain than the other, generating different habitats on different sides. So, if you're on the northeast side of Mauna Loa, for example, you're in very wet rain forest, whereas on the other side you have the Ka'ū Desert, with just nine inches of rain a year."

I feel slightly skeptical that the world's diversity of fruit flies should level out at some 4,200 species. It just seems improbable to me that Hawaii could turn out to be the only *Drosophila* megafactory on the planet. One certainty is that geneticists have a lot of raw materials to explore, should they decide to focus their attentions on other fruit fly species besides *D. melanogaster*.

When he's not studying the movement of flies on an archipelago, O'Grady is moving flies around the world. He runs the National Drosophila Species Stock Center (NDSSC), which resides in the basement of the Comstock building where we met. The NDSSC is the largest of three facilities that distribute different fruit fly species and strains to labs around the world, the others being in Japan and Austria. A much larger facility at Indiana University in Bloomington specializes in mutated strains of one species, *D. melanogaster*; it houses about 50,000 mutant lines and makes about 3,000 shipments per week to labs worldwide.

I asked O'Grady if I could see the NDSSC. We walked to the basement floor, where I was led into a home-kitchen-size room whose walls were lined with metal racks, each holding bins of vials held together in bundles of about ten. It was a much smaller and more modest affair than I had envisioned, but then fruit flies are very small creatures that do not demand spacious living quarters. The original shipment from San Diego numbered 750,000 live flies, and the room we were standing in held about a million.

It speaks to the adaptability of fruit flies that in the depauperate universe of a small clear plastic tube capped with a cotton ball, they are willing and able to carry out their entire life history, from courting and mating to egg laying, larval feeding, pupation, and emergence into the next generation of adults. As I had seen at Kelly Dyer's lab, some of these vials contain all stages at once. The yeasty mush at the bottom is mainly a mixture of sugar and nutritional yeast; some are supplemented with mushed banana, cactus fruit powder, and high-protein cereal. O'Grady pulled out a vial and pointed to tiny tunnels in the food medium, tracks left by munching maggots.

As described on its Facebook page, the NDSSC "currently maintains a living collection of 250 *Drosophila* species represented by 1,500 stocks that are used by biological researchers focusing on questions in evolution, ecology, developmental biology, physiology, neurobiology, comparative genomics, and immunology." Here, "stocks" refer to geographic populations of the same species. "We have a total of about 1,500 stocks," O'Grady reported. "Some species only have one stock, others more than a hundred.

About 40 *Drosophila* species have had their entire genome sequenced, and these species are in high demand. As soon as someone sequences a new genome, that fly can leap from obscurity to celebrity status."

The NDSSC was not always based in Ithaca. In fact, its history is about as stable as a Hawaiian island. Originating in Austin, Texas, in the 1930s, it moved to Bowling Green, Tucson, and San Diego before alighting in Ithaca in the fall of 2017. Business is steady, about a thousand shipments going out per year to all corners of the globe. Each shipment includes two or three vials of about 50 flies each, at a flat fee of $40 per species.

The biggest problem with customers is when flies arrive dead, in which case they are replaced with a new shipment. That happens maybe three or four times per year. Mishaps were more common in the first year following the move to Ithaca, where winter temperatures can quickly kill exposed flies. Heat packs are now used, with additional steps taken to prevent shipments languishing in cold or hot areas. Occasional losses are usually the result of bad luck, as when a shipment gets diverted on an airport runway.

It was strange to think that this modest room housed fruit fly lineages representing the entire swath of *Drosophila* evolution over its 60-million-year history. That's roughly the equivalent of the evolutionary scope of the entire primate order of mammals to which we belong.

Giant Sperm

We unintentionally aid the global spread of fruit fly genes when we ship them long distances, for there are inevitably escapees. More typically, flies and other animals disseminate their genes through sex. But mate choice isn't the end of the procreation game for flies, or many other animals; as we saw in chapter 8, both males and females may employ tactics that can favor one male's sperm over another. In a few fruit fly species, males have taken a strange route to improving their stakes in the genetic lottery.

First, a brief mention of prevailing theory. In promiscuous species, whose females will mate with several males, nature has tended to favor the production of large numbers of very small sperm cells. The evolutionary logic of this so-called sperm competition is straightforward: if you purchase more lottery tickets, you are more likely to pick a winner, or in this case, the odds are greater that one of your sperm cells will be the first (and last) to reach an egg. Contrasting mating systems help explain why male chimpanzees have much larger testicles than males of their larger cousins, gorillas. Chimpanzee mating systems are strongly promiscuous, so it pays to enter more sperm cells into the lottery. By contrast, in gorilla society, the dominant male has exclusive mating access to the females in the group; his monopoly precludes the need to produce large numbers of sperm. (Behaviorally and anatomically, humans lie between these two extremes.) Various forms of sperm competition have been documented widely among animals—from rodents to snakes to dragonflies.

Curiously, some fruit flies have taken a different path to gaining precedence over a competitor's sperm. I had the opportunity to explore this a little further during my conversations with Kelly Dyer.

"Flies' sperm varies a lot in size," Dyer told me. "The smallest is perhaps half a millimeter, which is itself gargantuan compared to human sperm. And the grand prize for longest fly sperm—indeed, the longest known amongst all animals—goes to *Drosophila bifurca*, which have sperm cells on the order of 6 to 7 centimeters."

What?! Recall that fruit flies are very small. This is like a human sperm measuring the length of a tennis court. By necessity almost all of that length is the tail, or flagellum, whose motion propels the sperm head.

How and why does it work? It turns out that longer sperm are really good at displacing their competitors from the female reproductive tract, which gives them an advantage in the competition for fertilization. Those outlandish tails form what Dyer describes as "a tangled mass of yarn" that impedes the progress of any subsequent male's sperm that may arrive on the scene.

This raises a rather glaring practical issue. How does a female accommodate these unwieldy sperm?

The glib answer to that question is that evolution takes care of it. When it comes to things that must interact intimately—such as the bits and pieces involved in sexual intercourse—evolution isn't going to sit idly by while one gender's equipment runs up an adaptive hill.

To provide an illustration, Dyer pulled an old book, *Evolution in the Genus Drosophila* (published in 1952), from her shelf and showed me detailed drawings of the reproductive tracts of female flies.

Dyer pointed to a drawing of *D. pseudoobscura*, a species with short-tailed sperm, showing female sperm storage organs of ordinary dimensions, for a fruit fly. Turning the pages, we beheld the reproductive tract of a species with giant sperm; these storage organs formed long coils like on an old telephone cord.

"What happens is that the female reproductive tract has evolved to accommodate the sperm of her mate. There's a very tight within-species correlation between sperm length and the length of her sperm storage organ."

"How much of the body cavity are these organs taking up?" I asked.

"Almost all of the abdomen."

Dyer led me into her adjoining lab to see the real thing. She knocked out a cluster of flies with carbon dioxide gas, then, using a pair of very fine needles mounted in cork handles, separated a female fly from the bunch and, in a small puddle of water, teased out the reproductive organs. Placing the sample on a microscope slide and thence onto the viewing platform of a dissecting microscope, Dyer invited me to look.

Each of two pearly white ovaries contained an orderly cluster of segments the pallor of a peeled lychee with a shape reminiscent of a mosque's dome. Nearby lay a pair of bumpy tubes resembling a string of pearls: the seminal receptacles. A dissected male yielded two yellowy spiral structures, the testes. Dissecting one of these exposed the infamously long sperm cells, which under this scope looked like a milky cloud. In that cloud lay many thousands of long-tailed sperm, a thousand times fewer than the much smaller sperm cells of a human ejaculate, and due to their microscopic slenderness far too small to see individually at this magnification.

Producing giant sperm is an unconventional reproductive strategy, but it clearly has worked for some flies. It's tempting to think that this is an example of males exerting pressure on females, but it appears the females are in charge. "Crazy sperm forms are evolving because female reproductive tracts are evolving that bias fertilization in favor of these weird, specific traits," says Scott Pitnick, who studies the giant-sperm phenomenon at Syracuse University. His studies show that female flies are proactively evolving larger sperm storage organs, and that males are adapting in response. Females with longer storage organs tend to remate sooner with other males, which further stokes competition among sperm and hence the advantage to males producing longer sperm. Lastly, because producing large sperm is energetically expensive for males, males producing more of them are likely to win the competition. This benefits the females' own investment by ensuring that they are fertilized by the fittest, most robust males. It goes to show that not all sexual selection happens where we can see it.

But what if the flies themselves cannot see what's going on around them? In about 1909, Fernandus Payne, a graduate student at Columbia University, condemned a population of flies to complete darkness for 49 generations. Why? He was hoping to demonstrate support for the theory of inheritance according to biological need developed a century earlier by the French biologist Jean-Baptiste Lamarck. Lamarck's idea, that the environment gives rise to changes in animals, predicts that living in darkness would

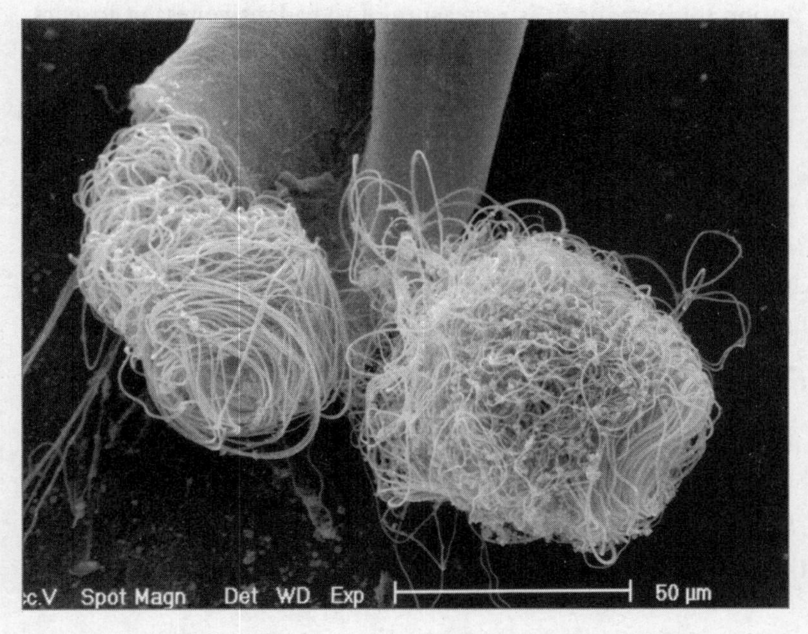

The extraordinarily long tails of certain fruit flies' sperm—just two shown here—make them the largest sperm cells of any known organism.

(© ROMANO DALLAI)

manifest in a measurable degeneration of the insects' eyes. But two years later, when the last generation of flies emerged bleary-eyed from the darkness and saw light for the first time, their eyes were as good as those of their ancestors 48 generations earlier. Sometimes fruit flies are as important for undermining a hypothesis as for supporting one.

That wasn't the end of the tunnel for the flies. A half century later, Japanese scientists decided to explore more fully the possible effects of living in the dark. Dark-fly is a *Drosophila melanogaster* line started at Kyoto University in 1954 and maintained in constant darkness for more than 60 years. With their generation span of about two weeks in ideal conditions, that amounts to about 1,500 generations, or the equivalent of 27,000 years of evolution for humans. It is a measure of the adaptiveness of insects, with their relatively short generation spans, that Dark-flies already show a reproductive advantage over their wild-type (non-light-deprived) counterparts in a mixed population under dark conditions. When female Dark-flies were bred with male wild types, they produced fewer young in dark conditions than when both parents were Dark-flies. It isn't yet known why, but scientists speculate that Dark-fly females might prefer Dark-fly males as partners, and that they may be using odors or sounds to discriminate partners in the absence of visual cues.

In time, maybe Dark-flies will reveal greater secrets of what it is and isn't like to be a fly. What seems quite certain is that flies, and fruit flies especially, will continue to be at the forefront of discovery in the fields of genetics and evolution.

Chapter 10

Vectors and Pests

> In the time it takes you to read this sentence, someone somewhere will be killed by a fly-borne disease and hundreds of others will become sick because of fly-associated microorganisms.
>
> —STEPHEN A. MARSHALL

Flies' diversity and opportunism have rendered them both beneficial and detrimental to human interests. As we've just seen, flies have played a major role in advancing our understanding of genes, behavior, and evolution. Now we turn to a darker side of flies' relations with us: their dual impacts as vectors of deadly and debilitating diseases and as agricultural pests.

It is hard to overstate the importance of flies in the course of human history. In his celebrated 1997 book *Guns, Germs, and Steel*—whose 20th-anniversary edition's cover features two bullets, a mosquito, and a steel nut—biogeographer Jared Diamond theorized that diseases mostly transmitted by flies were key players in the global expansion and lamentable colonization of Africa and the Americas by European nations. In *The Mosquito* (2019), Timothy Winegard, a political scientist and military historian at Colorado Mesa University, details how mosquitoes have protected Rome from foreign invasion since antiquity. They even protected Rome from itself, having essentially orchestrated the fall of the Roman Empire. Flies are credited with driving the Mongols from Europe once and for all in 1242, and their depredations on poorly prepared British troops determined the fate of the United States of America.

Mosquitoes transmit disease in various ways. Malaria parasites attach themselves to the gut of a female mosquito and enter a host as she feeds. As with yellow fever and dengue, the pathogen can also enter the mosquito as she feeds on an already infected human, and is delivered to a subsequent victim via the mosquito's saliva. Thus, mosquitoes act as flying contaminated needles, extending an ill person's infectious range for miles.

Not that mosquitoes have acted alone. Biting midges (Ceratopogonidae) are known to transmit at least 66 viruses, 15 protozoans, and 26 filarial nematodes. Oroya fever, transmitted by sand flies of the *Phlebotomus* genus, was responsible for killing about a quarter of Pizarro's army during the mid-16th-century conquest of the Incas. The tsetse fly slowed the march of European colonization of Africa by crippling the importation and use of horses and other draft animals via trypanosomiasis, a parasite-borne illness that causes sleeping sickness in humans, as well as nagana—an African disease of cattle and other hoofed mammals characterized by fever, lethargy, and swelling.

The Biggest Killer

For sheer havoc, however, mosquitoes take first place. A few hundred species of mosquito combined outrank humans for causing human deaths. Since the year 2000, mosquitoes have accounted for about 2 million human deaths per year, compared with 475,000 per year at the hands of our fellow humans. Researchers have credited mosquitoes with almost half the deaths in human history. That's about 52 billion out of 108 billion of us, by Winegard's calculation. Until World War II, many more soldiers died from infectious diseases spread by insects than as a result of injury on the battlefield.

For every human who dies, there are many more stricken by illness. Depending on whom you ask, mosquitoes alone are today responsible for sickening between 200 million and over 300 million people per year.

In all, over 15 diseases are transmitted to humans from mosquitoes. These pathogens form three distinct groupings: viruses, worms, and protozoans (single-celled microscopic organisms of the kingdom Protista). Malaria is caused by protozoans and carried by mosquitoes of about 70 of the 480 species of the genus *Anopheles*. *Aedes* mosquitoes, such as the Asian tiger mosquito, which because of us has spread to Africa and the Americas, carry the viral diseases yellow fever, dengue, chikungunya, Zika, and a half-dozen types of encephalitis. *Culex* mosquitoes are also culpable for encephalitis, as well as West Nile virus, and the filariasis and elephantiasis caused by nematode worms. Fortunately, there is no evidence of mosquitoes, houseflies, or other flies playing any role in the transmission of the coronavirus we're all thinking about these days.

Of all of these ailments, malaria (from the Italian for "bad air") is by far the gravest. The World Health Organization (WHO) reports that malaria was the sixth leading cause of death in low-income countries for 2016, accounting for just under 40 deaths per 100,000 population.

It was not until the 1890s that the connection between mosquitoes and malaria was established. Symptoms, which usually appear between one and four weeks but can take up to a year to develop, include fever, diarrhea, headache, sweats or chills, nausea and vomiting, and muscle and stomach pain. Left untreated, malaria can progress to severe illness, leading to coma, seizures, respiratory failure, and death.

The family of malaria parasites is among the planet's most devious manipulators and infiltrators of other life forms. Mosquitoes are its primary puppets, and humans its most inviting target, but we are not malaria's only quarry. There are over 200 types of malaria parasite vexing birds, bats, monkeys, antelopes, and lizards worldwide; only five of them afflict *Homo sapiens*. While their behavior and reproductive success may be affected, most animals have coevolved to cope with the parasite. One notable exception is native Hawaiian birds, whose exposure to introduced malaria is thought to have contributed to their declines and extinctions. The two most dangerous types of malaria to us are *Plasmodium falciparum* and *P. vivax*.

When it comes into contact with red blood cells, the malaria parasite produces a chemical lure that makes an infected host more attractive to the next mosquito that happens along, thereby significantly boosting the disease's spread. The parasite further manipulates a mosquito to bite more frequently by suppressing her output of anticoagulant; this limits her blood intake per bite, and she is compelled to bite again and again, giving the parasite more opportunity to disembark into the bitten host.

Malaria has a historical range much larger than its current one. In addition to the tropics, the disease once stretched as far north as Canada and northern Europe. When Europeans colonized the Americas, they did so with more than Europeans; they also brought a tidal wave of mosquitoes delivering disease and death against the nonimmune indigenous peoples. Prior to this time, while there were hordes of mosquitoes native to the Americas, they did not carry the lethal diseases that the newly introduced *Anopheles* and *Aedes* species did. As recently as 1935, roughly 130,000 Americans were infected by malaria, leading to 4,000 deaths. By 1950, through a combination of DDT spraying, wetland drainage, and removal of mosquito breeding sites, the disease was virtually eradicated from these regions. Today, 85 percent of malaria cases occur in sub-Saharan Africa, 8 percent in Southeast Asia, 5 percent in the Eastern Mediterranean, 1 percent in the Western Pacific, and 0.5 percent in the Americas.

As tropical disease vectors, flies also visited punishment on slaving nations. By importing its mosquito carriers, yellow fever afflicted more those countries that engaged in the slave trade than those that didn't. Starting in 1648, outbreaks took many lives in the West Indies. The famous slave uprising on the ship *Amistad* may owe its success to mosquitoes whose bites had weakened the crew with yellow fever but to which the slaves were mostly immune. Between 1693 and 1905, some 100,000 to 150,000 Americans succumbed to yellow fever.

You may wonder why these pathogens evolved to kill their sheltering hosts. The answer is that it is the symptoms of illness preceding death that get the job done for the pathogen. After that we are expendable. Depending on the ailment we are afflicted with, those symptoms could be coughing and sneezing, lesions, or open sores—all effective routes of transmission. Then there are our interactions with our fellow humans, including sexual ones, and our contact with other contaminated objects. The global coronavirus pandemic that has made social distancing, wearing facemasks, and frequent handwashing a part of everyday life has also made us more aware of the crafty means by which a pathogen can aid its own spread.

Given malaria's legacy and the mosquito's role as courier-in-chief, we may wonder little that the mosquito's ecological role has been characterized "as a countermeasure against uncontrolled human population growth." In fairness to the mosquito, theirs is an indirect culpability, for they are the vectors and not the direct agents of death. But it amounts to the same.

Counterattacks

What to do about so effective a menace? A major prong of modern attempts to control fly-borne diseases has been to hammer the vector species and its haunts with insecticide chemicals. This approach has the satisfying effect of measurably reducing enemy numbers and disease incidence, at least for a while, but it has drawbacks. Chief among them is how to reach the target organism while minimizing collateral harm to non-target species. It costs money, and there are logistical hurdles, such as needing to periodically reimpregnate pesticide-treated netting and other materials. Pesticides are also hazardous to us; they cause about 200,000 human deaths per year according to the United Nations. Lastly, there is the ever-present specter of insecticide resistance.

Genetic strategies could circumvent most of these problems. The two most prominent genetic strategies are the sterile male technique, and the generation of pathogen-resistant transgenic flies using gene-editing techniques. A third technique, gene drives, promises to greatly accelerate the efficiency of the other two.

The sterile male technique (SMT) involves raising legions of males in captivity (females are destroyed), rendering them sterile (but still apparently healthy) by irradiation of the pupae, then releasing them into the wild. The idea here is that the more sterile males there are, the lower the proportion of wild matings that are fertile, thus reducing the size of the next generation. SMT is dependent on a high ratio of sterile insects to wild fertile insects, ideally more than 10 to 1. Depending on the species, this requires releasing millions or billions of sterile males to be effective. The reason males are usually targeted is that females tend to mate only once, whereas males will mate as often as they can, accelerating the technique's impact. This approach is much more feasible with insects than with larger animals because their small size and high fecundity make it possible to cultivate huge numbers of them in captivity in a relatively short time.

In 2015, scientists first used the new gene-editing technique CRISPR to modify the genome of the so-called yellow-fever mosquito, *Aedes aegypti*. Also in 2015, researchers injected manipulated genes into *A. aegypti* embryos and demonstrated that the CRISPR technique could induce different mutations. Considering that specific gene sequences can be and have been identified as playing a role in such varied physiological processes as metamorphosis, embryo development, and host-pathogen interactions, and that researchers have successfully created mosquito populations showing growth retardation, loss of ovarian function, and reduced egg hatching rates, it isn't hard to imagine the potential of these techniques to manipulate wild insect populations to our own ends. Scientists have identified and manipulated the genes of *A. aegypti* to create masculinized genetic females with nearly complete male genitalia. Because only female mosquitoes feed on blood and subsequently transmit pathogens, the use of gene splicing to convert female mosquitoes into harmless males has shown promise as a vector management strategy.

Another way that scientists have found to interfere with mosquito reproduction involves widespread, mostly parasitic bacteria called *Wolbachia*. These bacteria have a predilection for insects. It's estimated that as many as 70 percent of all insect species harbor *Wolbachia* in their bodies.

Although it was first identified in 1924, research on *Wolbachia* didn't intensify until after 1971, when it was discovered that *Culex* mosquito eggs were killed when the sperm of *Wolbachia*-infected males fertilized infection-free eggs. While these bacteria are ubiquitous in mature eggs, they are practically absent from mature sperm. Therefore, only infected females pass the infection on to their offspring, and males are a dead end, so the more the gender ratio is skewed toward females, the better the score for *Wolbachia*. A masterful manipulator, *Wolbachia* has evolved several ways to favor females over males. One is simply to kill male embryos, another is to induce female embryos to eat male embryos, and a third is to ensure that only females infected with *Wolbachia* can successfully mate with infected males. This lowers the reproductive success of uninfected females, thereby promoting *Wolbachia*'s prevalence.

Wolbachia also inhabits nematodes, including one that causes elephantiasis. In this case, the *Wolbachia* has evolved a mutualistic relationship with the nematode, so much so that if you kill the *Wolbachia*, you kill the nematode. It follows that a major avenue in elephantiasis research is in seeking ways to kill the *Wolbachia*.

More recently, a chance discovery was made that if a fly is infected with *Wolbachia*, its ability to transmit viruses is impaired. That got the attention of researchers of mosquito-borne viral diseases, most notably dengue. If mosquitoes infected with *Wolbachia* transmit less of the dengue pathogen, then perhaps *Wolbachia* can be used to reduce dengue infections. A trial in the northeastern Australian city of Townsville, with 187,000 inhabitants and a significant dengue presence, had no cases in four years following introduction of mosquitoes infected with *Wolbachia*. *Wolbachia* also shows potential for reducing threats posed by the chikungunya and West Nile viruses.

As Kelly Dyer told me, "The beauty of this approach is that you're not affecting the populations of the mosquito vector, which could have dire ecological consequences" (due to mosquitoes' key role in food webs). Dyer describes this area of research as "insanely active. Bill Gates, for one, has put a ton of money into *Wolbachia* research."

If bacteria can favor or eliminate certain mosquitoes in a population, might we manipulate the mosquitoes' genes to do the same thing? Enter the gene drive. Gene drives work by generating a superheritable gene that can quickly spread through the affected population. This points to an important difference between gene drives and traditional transgenic methods. While transgenic flies can reasonably be assumed harmless because natural selection will reliably eliminate any escaped mutant flies, the expectation with flies subjected to a gene drive experiment is the opposite: the manipulated organism will be favored by natural selection, at least in some fraction of the wild population.

It is at once an exciting and a sobering idea. This strategy could allow pests to be locally removed without affecting any other species or populations elsewhere. Gene drive systems could also directly reverse evolved insecticide resistance, giving once-effective compounds a new lease on life. Alternatively, so-called sensitizing drives might render resistant insects vulnerable to relatively benign compounds, perhaps even ones that are otherwise completely nontoxic to humans and the environment.

Another hope harbored by gene drive advocates is that pests might be altered so that they no longer consume crops but otherwise perform their natural ecological functions. What if we could, for example, manipulate a fly's olfactory system so that it was no longer attracted to the target crop?

On the grimmer side, by causing genetic changes that self-propagate, gene drives pose ecological risks by drastically altering the genetic makeup of wild populations. Indeed, theoretical models have shown that even the entry of a few individuals with gene drive constructs can lead to a complete invasion of a population. It is easy to imagine that gene drives could be a game-changer in the control of vector-borne diseases, and why they have sometimes been called extinction drives.

Gene drive advocates remind us that the approach is intended to *suppress* target populations, not make them extinct. They also point out the likelihood of natural resistance to these schemes. The introduction of a genetically manipulated population of insects would impose tremendous pressure favoring resistance mechanisms in the target population. These include natural selection for drive-resistant genes, preferential inbreeding, and even asexual reproduction in capable species. Over time, these sorts of mechanisms are not considered merely possible, but inevitable.

A recent risk workshop examining possible harms from using gene drives to control the malarial mosquito *Anopheles gambiae* concluded that, while risks exist, they are likely dwarfed by the impact of malaria. In a successful eradication scenario, there are many other local mosquito species likely to fill the vacated niche of the targeted malaria mosquito vector. Moreover, no predators or pollinated plants are known to rely primarily on any one mosquito species, so their predators could simply switch from one to another. Lastly, even if a gene drive were threatening to render a targeted mosquito extinct, scientists believe they could "rescue" it by re-editing its genome. A cost-benefit analysis—taking into consideration the target species, ecosystem, and nature of the change in question—is recommended for any proposed gene drive effort.

The enormous toll in the suffering and death caused by malaria and other mosquito-borne illnesses predicts that humans will be willing to take these risks. By September 2015, the Bill and Melinda Gates Foundation had granted $75 million to a project out of Imperial College London. As part of its Target Malaria effort, researchers there had engineered a gene drive to suppress laboratory populations of *A. gambiae*, the most important vector of malaria in sub-Saharan Africa.

The success or failure of new genetic techniques may lie more in the hands of nature than in our own. The high fecundity and short generation spans of insects renders them formidable opponents of whatever tactics we may challenge them with. Evolution fights back with tactics of its own that may hinder gene drives, including genetic variation in natural populations, evolved resistance through mutations selected for under gene drive pressure, and nonrandom breeding patterns. Experiments in multiple insect species have reported both the emergence of resistance to gene drives and the prevalence of natural genetic variation that thwarts the CRISPR mechanism from spreading genes throughout populations as intended.

Resistance

It's little wonder that genetic techniques are being so actively pursued in our quest to defeat mosquito-borne illnesses, for the history of insecticide use has been one of developed resistance in the target insect. Nowhere is that more apparent than in our efforts to defend ourselves against malaria.

In 1961 malaria cases in India had dipped to less than 150,000, down from 75 million in the early 1950s, with 800,000 dying in a single year. But DDT saturation campaigns—picture a nation doused with 60 million pounds of insecticide in one year—invited resistance. Malaria staged a comeback. India suffered a major epidemic in 1976, with roughly 25 million cases. In Indonesia, malaria cases quadrupled between 1965 and 1968. By then, the WHO had officially recognized the failure to eradicate the disease. By the early 1990s at least 100 species of mosquitoes and other disease-transmitting vectors were showing resistance to various insecticides. And because the embattled parasite was now more resistant than its precursors, new infections were particularly dangerous. In 2000, 10 percent of the world's population suffered from malaria.

Anti-malarial medicines reveal a similar pattern. First used to combat the disease in Rome in the early 17th century, quinine was no longer effective by the late 1940s, and was replaced by chloroquine. By the 1960s, chloroquine was useless in most of Southeast Asia, South America, India, and Africa. Resistance to its successor, mefloquine, was confirmed only one year after its commercial release in 1975.

Overall resistance to pesticides rose 61 percent between 2000 and 2010, according to Bruce Tabashnik of the University of Arizona. The WHO's report for 2010 to 2016 shows that resistance to the four commonly used insecticide classes—pyrethroids, organochlorines, carbamates, and organophosphates—is widespread in all major malaria vectors throughout Africa, the Americas, Southeast Asia, the Eastern Mediterranean, and the Western Pacific. A 2016 clinical trial of Mosquirix, a long-awaited malaria vaccine, involving 447 African children aged 5 to 17 months, yielded disappointing results. Overall efficacy across seven years of follow-up was 4.4 percent, dropping close to zero in the fourth year, with possibly net negative effects thereafter. However, in a larger trial published in 2015, involving 15,459 infants and young children in seven African countries, Mosquirix reduced the number of cases of clinical malaria by 39 percent. On the basis of evidence to date, the WHO and certain other medical agencies believe that the benefits of Mosquirix outweigh the risks of adverse reactions, such as meningitis and convulsive seizures.

During this time period, one class of insecticides, pyrethroids—a group of man-made pesticides similar to the natural pesticide pyrethrum, which is produced by chrysanthemum flowers—have dominated vector control efforts. That presents a problem, because the history of mosquito vector control shows the danger of relying on a single class of insecticide. As we've seen, if you continue to bombard an organism—especially an abundant, fast-breeding insect—with the same weapon, resistance is almost bound to follow. Sure enough, recent years have witnessed an alarming increase in pyrethrin-resistant mosquito populations.

Sometimes the most reliable methods are the most basic. Insecticide-treated bed nets limit exposure and severity of disease through fewer bites. Between 2000 and 2016, malaria cases dropped 40 percent worldwide, which amounts to about 663 million fewer cases. More than two thirds of this decline is credited to the use of bed nets treated with long-lasting insecticide, with another 19 percent attributed to indoor wall spraying.

A conversation I had with Priscilla Tamioso, a biologist who works for the Dengue Control Program in the southern Brazilian city of Florianópolis, underscores the importance of simple, practical measures. This far south, the threat of malaria gives way to dengue, Zika, and chikungunya.

"Our focus is public education," Tamioso told me. "We visit different neighborhoods looking for mosquito breeding reservoirs. We also sit down with residents to explain the importance of reducing such reservoirs. Southern Brazil receives a lot of rainfall, so for example, drilling holes in abandoned car tires, or covering them up, is important."

As it is for evolved resistance to insecticides, the more effective the defense, the more concerted the counterattack. Some mosquito species are adapting to bed nets by becoming resistant to the pyrethroid pesticides with which they are infused and by shifting their feeding schedule from nighttime to daytime.

Dengue is the most important reemerging insect-vectored viral disease in the tropics and subtropics. Only nine countries had had severe dengue outbreaks before 1970; today, according to the WHO, the disease is endemic to 100 countries. In October 2019, this normally flulike but sometimes lethal fever hit Nepal, a nation previously too chilly to have worried about the disease. Within two months, at least 9,000 people were ill, and 6 had died. Global heating in the high-altitude nation—which saw its first-ever dengue cases in 2006—is leading to longer periods of tolerable temperatures for the *Aedes* mosquitoes that serve as vectors, as well as bigger monsoons, which multiply water reservoirs suitable for breeding. In 2019, the Americas also saw a record number of dengue cases: 2.7 million.

Whether or not we can ever free ourselves from these diseases, we can at least improve our ability to detect them. An Australian research team led by Dagmar Meyer of James Cook University in Cairns has developed a mosquito trap to detect the presence of disease-causing viruses circulating in the wild. They remodeled an innovative device launched in 2010 that lures mosquitoes into tasting honey-coated cards, then monitors saliva left at the scene. As you can imagine, that's a minuscule volume of fly drool: about five billionths of a liter. Volumes like that test the limits of current detection systems.

But mosquitoes leave behind about three thousand times more urine—recall these biters' adaptation of expelling water to concentrate the blood while the mosquito tanks up. The remodeled traps, using a pee-collecting card, use standard overnight light traps and longer-standing traps that exhale delicious carbon dioxide to lure the bugs in. When a mosquito enters a urine trap, her excretions drip through a mesh floor onto the card. The researchers used 29 urine traps alongside saliva traps placed in two insect-rich spots in Queensland.

The urine traps detected genetic traces of three pathogens—the viruses that cause West Nile, Ross River, and Murray Valley encephalitis—while the saliva traps detected two. Among its advantages, the card-based method avoids the constant refrigeration that checking whole mosquitoes requires, and it's not as labor intensive or cruel as the old method of exposing sentinel chickens or pigs and monitoring them for signs of infection. The test is being touted as an early warning of disease risk from local mosquitoes.

Cards are not the only method being developed for monitoring dangerous hitchhikers in mosquito excretions. Researchers from the UK and the USA in 2017 presented a highly water-resistant cone from which could be collected mosquito pee and poo. They were able to detect the DNA presence of filarial worms, malaria plasmodia, and flatworms in artificially exposed mosquitoes.

In the wake of earlier eradication efforts, cold winters and shorter warm seasons have kept malaria and other fly-borne diseases out of more northerly regions of the world, such as the northern United States, Canada, and Europe, but that situation may change as global temperatures rise under the influence of human activities and as international travel and trade resume after the coronavirus pandemic. European clinics and hospitals are now treating eight times more malaria patients than in the 1970s, and malaria rates in Central Asia and the Middle East are up tenfold. In addition to dengue, leishmaniasis and encephalitis are on the rise, threatening to spread into many areas of Europe.

Farm Wars

Understandably, flies garner more angst from us for their potential to kill us by transmitted disease than for their role as pests of our food supplies. But flies have not missed the opportunity to capitalize on our massive crops, especially fruits, or on our passion for raising and consuming livestock.

First off, we need to recognize that *pest* in this context is an anthropocentric term based on context and human values. We use "pest" like we use "weed," to designate something strictly in human terms, without regard for its ecological value beyond us. Bees pollinating flowers in someone's landscaped garden would not be regarded as pests until they nest under a porch eave, nor would flies decomposing a dead rodent be pests, but those laying eggs in a fruit crop would be. To us, these contexts are viewed as quite different. Beyond the human context, they are simply examples of the critical roles insects play in cycling nutrients through ecosystems.

Only about 1 percent of insect species are considered to be of any negative economic importance, but that small proportion makes a big impact. Depending on what source you consult, between 15 and 50 percent of all food grown for human consumption is lost to damage caused by insects. Flies are significant contributors, but they do not make the top 10 list from London's world-renowned Kew Gardens, which features moth larvae, whiteflies (not true flies), spider mites, flower beetles, and aphids.

Any significant pest species must be abundant, and one avenue to hyperabundance in an insect population is to provide it with large swaths of land dedicated to growing its favorite food, and a dearth of natural predators or parasites. Therein lies a central dilemma of monocrop agriculture. A vast field of corn or a sprawling apple orchard is easier to cultivate and harvest than mixed-crop acreage, but it also presents an ideal opportunity for so-called pest organisms to flourish. When we plant acres and acres of corn, we may complain mightily, but should we really be surprised when the corn earworm and European corn borer and other insects indulge in their favorite (sometimes only) host plant?

The most significant fly crop pests are fruit flies. Most types of fruit flies evolved to feed on what has historically been a relatively rare resource, at least in temperate zones—fallen fruits that have begun to ferment and are of no use to the fruit grower. It doesn't take much force to penetrate the skin of a rotting fruit, which is usually ruptured anyway.

But there is a troublesome exception to this pattern. The spotted-winged fruit fly, *Drosophila suzukii*, is a serious pest because it is not a picky eater and is one of the few *Drosophila* species able to inject its eggs into the more resistant skins of fresh fruit. During my visit to Kelly Dyer's fruit fly lab, I had the opportunity to see these specialized flies up close. Peering through a dissecting microscope, I saw that, by fruit fly standards, its ovipositor is huge, and it looks like a serrated knife. As fruit fly specialist Martin Hauser with the California Department of Food and Agriculture explained to me, this high-tech apparatus includes sensory hairs (setae) at the tip that probably "taste" the fruit and monitor depth of penetration, and a barblike structure that probably anchors the ovipositor, preventing it from slipping out of the fruit when an egg emerges. It's worth bearing in mind that these structures must also accommodate the complex genitals of a male during copulation.

D. suzukii is not native to the United States, having arrived from Southeast Asia in the last ten or so years. These flies have developed a particular fondness for Georgia, where the blueberry industry is huge. The annual cost to American fruit growers of this fly's presence has already reached $400 million. The species is also spreading fast along the seacoast of southern Europe, dining on a convenient buffet of citrus, figs, cherries, and blackberries.

What are we doing about it? Classical biological control refers to the introduction of a predator or parasite that targets the pest

The specialized ovipositor of the spotted-winged fruit fly (*Drosophila suzukii*) allows her to penetrate and lay eggs under the skin of fruit.

(© MARTIN HAUSER, CALIFORNIA DEPARTMENT OF FOOD AND AGRICULTURE)

species. Tachinid flies represent a broad reservoir of parasitoids that target plant-eating insects (flies included), and for that reason they are widely effective in controlling insect pests of plants. Several species of these flies have been moved around the world to suppress crop pests, such as the winter moth in North America by introduced flies from Europe. This approach has the time- and cost-saving advantage of requiring little or no further intervention. In a 2006 study, entomologists Mace Vaughan and John Losey estimated that pest-controlling insects saved $4.5 billion a year in the United States alone.

However, introductions are always risky, and there have been many disastrous outcomes. A tachinid fly introduced to North America from Europe in 1906 to help control another non-native, the gypsy moth, has expanded its menu of over 200 known hosts to include many benign native species whose populations have since suffered precipitous declines. These include two gorgeous giants, cecropia and luna moths. The fly is killing about 80 percent of cecropias in Massachusetts, according to one study.

Other methods have their own drawbacks. The three big problems of insecticides are toxicity to humans, toxicity to nontarget organisms, and resistance. When we bombard an organism with something intended to kill it, be it a chemical or a parasite, resistant mutations and adaptations are immediately favored in the population, and the more evolutionarily agile the target organism, the sooner resistance is likely to appear. Insects, with their ability to produce multitudes of young in a short time, are such organisms.

The most effective response is a multipronged strategy. You may have encountered its name: integrated pest management (IPM). Like a book of chess openings, IPM encompasses a wide range of tactics. It also aims to avoid the ravages of chemicals that kill indiscriminately. IPM's methods include mass trapping of adult flies using attractants; protecting crops with exclusion nets; growing crops in enclosed tunnels; spraying with natural repellents; harvesting more frequently; postharvest refrigeration, fumigation, or irradiation; and the introduction of (preferably native) parasitoids, general predators, or insect-killing fungi.

Research into eco-friendly natural pesticides, which is ongoing, exploits plants' natural adaptations against defoliation and invasion of their tissues by insects. In tests, tea tree oil, andiroba oil, and citronella have each caused 100 percent mortality to houseflies and/or horn flies, and an application of hemp oil to organic crops proved highly toxic to houseflies and aphids. Perhaps because they are less profitable than commercially produced insecticides, relatively little is known about the mode of action of natural compounds—for example, whether they are neurotoxic or whether they simply suffocate the flies.

Then there are compounds that use the insect's sensory systems and behavior to lead it astray. A 2019 study found that both naturally and synthetically derived blends of mating pheromones render male swede midges—a serious pest of swede (rutabaga) crops—unable to locate females.

Asian farmers have been keeping fruit flies at bay for 70 years or more using the simple, if rather labor intensive, technique of bagging the fruit in situ. Yields of such crops as mangoes, melons, and cucumbers rose between 40 and 58 percent using this technique. Malaysia's carambola (star fruit) export industry, which was already worth US$10 million back in 1994, protects entire orchards by bagging the fruit on the trees.

Natural approaches tend to cause less collateral damage, both in terms of harm to other species and persistence in the environment. While those hemp oil applications to organic crops were toxic to houseflies, beneficial invertebrates like ladybugs and earthworms were unaffected. In contrast, ivermectin, a drug administered orally to livestock to rid them of internal parasites, such as botflies, does not break down completely in the bovine gut. Excreted in manure, ivermectin residues can persist for 20 years or more, rendering the dung a killing field for beetles, flies, and other beneficial dung fauna.

In 1980, the discovery in California of the Mediterranean fruit fly, or medfly, an illustrious invader of soft-skinned fruit and vegetable crops, led to a controversial aerial spraying campaign that blanketed 1,400 square miles with the insecticide malathion. Encompassing a 43-city area populated by two million residents and costing a good chunk of the $100 million the state threw at the eradication effort, the campaign drew the ire of many residents, who were advised to stay inside with their pets. In the current era of global commerce, the flies have since returned occasionally to the region, but they have been held in check, without any spraying, by the release of sterile males.

In combination with malathion, the sterile male technique (SMT) is credited with halting, and to some extent reversing, the expansion of the medfly into Central America and southern Mexico. Arriving in Costa Rica in 1955, the species had reached southern Mexico by 1979. A concerted SMT campaign, the ongoing Moscamed program, was launched in the late 1970s. Funded primarily by the United States and Mexico, Moscamed illustrates the colossal scale of sterile-male production deemed necessary to make such a campaign effective against so prolific a target. For much of the campaign, four breeding facilities in Mexico and Guatemala have been churning out well over half a billion flies per week. Between 1979 and 2016, 1.52 *trillion* sterile male medflies were reared. The fly has been eradicated from approximately 1.42 million hectares (over 500,000 square miles) and kept out of most of Mexico and the United States. There have been inevitable setbacks, usually triggered by El Niño weather events, resulting in rapid fly population increases referred to by insiders as fly storms. Nevertheless, with horticultural industries in the region valued at millions of US dollars per year, and tens of thousands of new rural jobs created, Moscamed's price tag of roughly $1 billion to date is regarded as an extremely good investment, with an estimated cost-benefit ratio of 150 to 1.

One of SMT's most notable successes is an ongoing campaign begun in the late 1950s that led to the eradication of the New World screwworm from the United States, Mexico, and Central America north of the Panama Canal. Screwworm larvae feed on the living tissue of their livestock hosts, gaining access through tiny wounds in the manner of botfly larvae. Less considerate than their botfly cousins, however, screwworm larvae can cause wounds and infections lethal to the host.

Dave Taylor, an entomologist with the USDA, explained to me that the biology of the screwworm made it uniquely susceptible to SMT. In contrast to medflies, screwworms naturally occur at relatively low population levels in the environment—as low as 5 to 10 flies per square kilometer by some estimates—so reaching quota is not a daunting objective. Most other insect pests are present in the environment at much higher populations. For instance, stable flies, which Taylor works on, can occur at levels of tens to hundreds of thousands per square kilometer.

"Furthermore," Taylor told me, "screwworm adults are completely innocuous. They do no direct damage to any commodity. It was therefore possible to release large numbers of adult screwworms with little notice from the community. It will be very difficult to convince the public to let you release hundreds of thousands to millions of [biting and/or disease-carrying] flies in their districts."

In the modern era, even a successful campaign may be only temporary. With expanded international transport, globalization of economies, and rapid long-distance movement of livestock and animal products, reinvasion is a constant threat. As of 2017, screwworms have been confirmed present in Florida. The species is still found in parts of Central and South America, and it has an Old World counterpart in parts of Africa, Asia, and the Middle East.

Biting flies combine to make up the most damaging arthropod pests of cattle worldwide. Two hundred thousand stable flies emerge from an average-size winter hay feeding site, and their collective blood feeding efforts reduce annual cattle milk production and weight gain for a total estimated cost to US cattle industries of $2.2 billion per year. (This is the species shown in the photo insert, in the before and after shots of a stable fly that bit my leg.)

With the benefit of hindsight, we might acknowledge that our reliance on chemical insecticides has done more harm than good, even creating new pests by suppressing competitors. As Gregory Paulson and Eric Eaton point out in their 2018 book *Insects Did It First*, we sprayed the boll weevil into submission, only to have the tobacco budworm take its place. For at least two good reasons, insects adapt to our chemical weaponry. For one, their short generation spans accelerate the rise of effective mutations and subsequent resistance. Second, insects, especially those strongly codependent on plants, get a lot of practice responding to defensive chemicals produced by the plants themselves. They either learn to cope with them, or they actively detoxify them. A further drawback to pesticides is that they tend to wreak more havoc against beneficial predatory and parasitic insects (many of which are flies) than against the intended targets. Reducing pesticide use and impacts is one of the three flagship projects of the US-based Xerces Society, an international organization dedicated solely to the conservation of all invertebrates, including insects.

In an era when biodiversity is already shrinking at an unprecedented rate due to human activities, I wonder if we would do better to deindustrialize food production methods. If you have ever seen a vast monoculture of grain stretching to the horizon, then you've seen an ecological desert and a bonanza to specialized insect pests. Alongside the clearing of forests to create grazing land for cattle ranchers, the growing of grains to feed livestock represents a deep-rooted inefficiency in how we feed ourselves.*

*As of early October 2020, more than 32,000 fires were burning in the Amazon rainforest, most of them started by ranchers to clear land for cattle grazing. Reuters, "Brazil's Amazon Rainforest Suffers Worst Fires in a Decade,"

As Michael Pollan explains in *The Omnivore's Dilemma*, when farmers grow a diversity of integrated foods—which they are encouraged to do when consumers buy local produce—they can give up most of their fertilizers and pesticides, because a diversified farm produces much of its own natural fertility and pest control.

Agriculture aside, flies' taste for blood, especially our own, combined with the serious health hazards stemming from their vampiric habits, generates enormous human angst with suffering on a large scale. As long as the human footprint on Earth continues to expand, we should not expect the fly footprint on humans to contract. With their prodigious numbers and their brief generation spans, insects are nimble opponents able to meet many of the challenges we throw at them. Mosquitoes are also remarkably good travelers. They can survive the frigid temperatures and low air pressures in the baggage compartments of transoceanic flights. By those standards, shipping containers, cars, and trains are a cinch. Combine these characteristics with changing climate patterns, and we may expect to be grappling with our mosquito foes for a long time yet.

Perhaps our technological ingenuity will outsmart flies, and

The Guardian, October 1, 2020, https://www.theguardian.com/environment/2020/oct/01/brazil-amazon-rainforest-worst-fires-in-decade. As another measure of inefficiency, consider that livestock takes up nearly 80 percent of global agricultural land, yet produces less than 20 percent of the world's supply of calories. Hannah Ritchie, "How Much of the World's Land Would We Need in Order to Feed the Global Population with the Average Diet of a Given Country?," *Our World in Data*, October 3, 2017, https://ourworldindata.org/agricultural-land-by-global-diets.

we will free ourselves from their depredations. That we should try is understandable given the suffering and death caused by malaria and other fly-borne illnesses. Nevertheless, it's a scenario fraught with its own dangers. We've known for half a century or more that removal of keystone species results in large-scale changes to the stability of an entire ecosystem. The ecological consequences of the loss of prolific and widespread fly species could be catastrophic. If a council of animals were held to adjudicate the future fate of biting flies, our votes for extermination would surely be opposed by those of the bats, birds, fishes, frogs, and other insects whose diets include them.

More likely, our efforts to eradicate fly disease vectors will succeed in temporarily suppressing but not eliminating their populations. That may be just as well. The folly of combative approaches to elements of nature we deem undesirable is that we lose sight of the interconnectedness of life. Rachel Carson did more than anyone to point out the sober ramifications: "We poison the gnats in a lake and the poison travels from link to link of the food chain and soon the birds of the lake margins become its victims."

Chapter 11

Detectives and Doctors

God in His wisdom made the fly
And then forgot to tell us why.

—OGDEN NASH

Flies' large role in epidemiology and agriculture shrouds two lesser-known benefits in the fields of forensics and medicine. So reliable is flies' ability to detect the aroma of a fresh human corpse that the appearance of eggs or live-born maggots and their subsequent growth and pupation schedules can be used as an accurate timetable. Forensic, or medicocriminal, entomologists collect the insects and, armed with an intimate knowledge of the life history of the fly species, can determine the time of death often to within an hour. This dipteran detective device has assisted hundreds of murder convictions.

There is a whole suite of larvae whose carrion-loving habits aid detective work. Among them: blowflies (Calliphoridae), flesh flies (Sarcophagidae), houseflies and their kin (Muscidae), soldier flies (Stratiomyidae), scuttle flies (Phoridae), and winter crane flies (Trichoceridae). The two most important of these for forensic entomology are the blowflies and the flesh flies whose janitorial talents we got acquainted with in chapter 6.

Due to their diminutive size, scuttle flies are able to find routes to buried bodies off-limits to the larger blowflies and flesh flies. Their teeny new larvae can penetrate the cracks and seams of most coffins, earning them the monikers coffin flies and mausoleum flies. Because of the slower rate of decomposition of a well-sealed body, several generations of coffin flies can colonize a single cadaver. Case in point: the presence of active maggots on a body buried for 18 years in Spain.

The key to flies' utility to forensics is their acute attraction to bodies in different degrees of decomposition, in different locations, at different times of year, and in different settings. An adult body lying in the open for a week in a Canadian woodlot in September is going to attract a different dipteran fauna from that attracted by a body buried for a month in a shallow grave in June outside São Paulo, Brazil. There's a range of insects that predictably colonize the body.

Although it generates the most societal interest, the role of insects in helping to solve suspicious deaths or homicides is but one of three branches of forensic entomology. In addition to *medicocriminal forensic entomology*, we have *urban entomology*, and *stored-product entomology*. Urban forensic entomology deals mainly with insects that interact with us in residential or commercial settings, such as legal cases involving termite damage. Disputes involving food contamination by insects, their body parts, and/or their poop fall under the domain of stored-product forensic entomology; an example would be the infestation of stored grain by weevils. Our attention here will focus on the medicocriminal branch, whose main protagonists are flies.

There are two broad domains of medicocriminal forensic entomology (hereafter just "forensic entomology"): the first, *initial colonization*, involves entirely Diptera, and mostly blowflies, which need fresh food and don't have the mouthparts to break down dried-up tissue. This branch concerns the developmental stages of insects colonizing a body during the first few weeks after death. It is enormously helpful that some flies can detect a decaying body within a few minutes of death. The second domain examines *successional colonization* of bodies in more advanced states of decomposition. As bodies rot, they go through stages of biological, chemical, and physical changes, each of which attracts different groups of insects. Other insects are adapted to eat only tougher tissue.

Not all flies that visit dead bodies come for the same reasons. Many come for a meal; blood and body fluids are a rich source of protein for nourishing egg development in females or semen production in males. Others come to breed. Some of these will have already mated elsewhere and are seeking a good place to deposit eggs or larvae. Yet other guests are drawn not to the corpse itself but to the fauna it attracts. Cluster flies, for instance, are parasitoids of earthworms. Females of the flesh fly *Sarcophaga utilis* (for which I could not find a common name) are parasitoids of dung beetles, but they will often find male suitors perched hopefully on or near carrion. Because eggs and maggots do not run or fly away when disturbed, it is the insects that breed in carrion that are most useful for estimating the postmortem interval, or PMI, the time elapsed since death.

PMI is a critical measure in the field of forensic entomology. Because flies, and particularly blowflies, are initiators of decomposition, they are the most accurate and most important indicators of time of death. Entomologists use knowledge of the current developmental stage of collected larvae, coupled with measurements of weather and temperature conditions, to estimate the PMI.

A variation on PMI is the minimum time since death, or minimum PMI (minPMI). MinPMI is estimated by identifying the oldest immature larval stages present on the body, estimating their age as indicated by their state of development, then, taking weather and other environmental conditions into account, working backward to estimate the date when the eggs or larvae were deposited. Sometimes the presence of a given species on the body can provide a critical clue, as when the body's location falls outside the species' normal range, indicating that the body has been moved.

Decomposition is accompanied by a procession of chemical smells. Early in decomposition, an array of microorganisms facilitates release of inorganic gases and sulfur-containing volatiles from the digestive and excretory systems. Thereafter, a variety of gases, liquids, and smelly organic compounds emanate from muscles, fat, organs, and other soft tissues. In all, hundreds of chemicals are released during decomposition, and there is much yet to learn about which ones stimulate carrion-attracted flies. As a body decays, its nutritional content changes, and these changes are reflected in emanating chemical odors. Flies and other carrion feeders cue in to these odor signatures and pay a visit only when the corpse is suitable for their specific needs.

By marking flies, Leo Braack (see the photograph on page 5) in 1981 discovered that flies could detect a rotting carcass from 40 miles away. They also can find their way to a lofty cadaver, as in the case of one colonized body 11 stories up in a Malaysian high-rise. Other flies will burrow as far as 6 feet down through soil to reach their decomposing target.

Nor are containers reliable barriers to certain flies. A study sought to determine whether and how soon flies might colonize dead humans inside suitcases—a tactic used by murderers to conceal victims. The flies were strongly attracted to the baits used (chicken livers, and one pig's head), and the smallest baby maggots were able to squeeze through the gaps between zipper teeth, an ability useful in solving some murder cases.

Often, human remains are not discovered until long after death, and this is when species attracted to the drier tissues of advanced decay come into play. Those cheese skippers we met in chapter 1 are late colonizers, and forensic entomologists have used the presence of their larvae to estimate time of death of bodies undiscovered a long time. Though they can appear on remains less than two months old in warm, humid locations like Florida, these flies may not appear on an exposed corpse until three to six months postmortem, usually after the body has completed the "active decay" decomposition stage and is beginning to dry out. Furthermore, as is not the case with some other insects used in forensic investigation, the presence of drugs, like heroin, does not significantly alter the development of late colonizers.

We may feel disgust at the thought of insects feeding on our dead bodies. How dare they?! But the insects are, of course, opportunistic bystanders. They know nothing of reverence or shame or the other strange customs of human civilization.

It should be added that forensic entomology is not restricted to cases involving death or humans. Prolonged abuse or neglect of young children, or of elderly or infirm patients, can lead to dead or dying tissues that draw the attention of flies normally drawn to a corpse. Forensic entomology methods also apply to nonhuman cases, such as animal abuse or neglect, or poaching. During a presentation at a Humane Canada conference I attended in Montreal, Dr. Margaret Doyle, a forensic veterinarian with Horizon Veterinary Group, said, "I wish we did more forensic entomology, because I really like maggots." She was referring to a case involving a cat named Snowball, who was found suffering from a maggot-infested wound on her hindquarters. The cat's "guardian" claimed that Snowball had had an accident the previous day, but when Doyle sent samples to Dr. Gail Anderson, the maggots were found to be third-instar larvae at least five days old, debunking the owner's account.

Expertise

To learn more about the field, I spoke with Anderson, a professor at Simon Fraser University and codirector of the Centre for Forensic Research. Gail got started in forensic entomology as a graduate student in the 1980s, when a professor interested in the field called her into his office.

"He asked, 'Gail, do you want to be a forensic entomologist?' I said, 'Cool! What's that?' So I took it on, and I never looked back."

Gail's path notwithstanding, the application of fly evidence to crime solving is not new. The first recorded case dates to 10th-century China. A woman claimed that her husband had died when their house burned down. But when his charred remains were inspected, traces of fly maggots were found on the back of his head. The man had lain dead for some time before the fire was started, and maggot traces marked the spot where he had received a fatal wound.

Anderson explained to me that forensic entomology grew more modern in the 1800s, especially in Germany and France. By the 1930s, Britain was getting on board, with a famous case in 1935 involving a physician named Buck Ruxton.

I looked up Ruxton. A doctor in private practice, he was convicted of murdering his wife, whom he suspected of having an affair. He also murdered their housemaid when she happened upon the scene. Having dismembered their bodies (with great skill) he disposed of them in a ravine. More than a week later, a keen-eyed pedestrian on a bridge spotted the remains, which had washed downstream. Some of the pieces were wrapped in newspapers, which helped pinpoint their upstream origin. The presence of 12- to 14-day-old maggots on the victims helped lead to the conviction, and hanging, of Ruxton.

In their textbook *The Science of Forensic Entomology*, David Rivers and Gregory Dahlem provide an illuminating metaphor for the conspicuousness of a decomposing body to corpse-seeking flies, which do not rely mainly on vision to find their meals: "Think of a dead body as a light with a dimmer switch. After death, the chemical signals show the body as a dim light in a dark environment. As time progresses the chemical signals intensify, making the body 'shine' brighter and brighter." As decay peaks and then wanes, the body becomes once again "dimmer" and less attractive to insect customers.

I asked Anderson how many types of flies colonize human bodies?

"About fifty species in North America. I deal regularly with about six to ten blowfly species. The species profile will differ if you are in Frankfurt or Mumbai, but within those regions the colonists are predictable. Diptera [flies] are the stars here; they are involved at all levels of decay. The only other significant visitors are some Coleoptera [beetles]."

The American Board of Forensic Entomology (ABFE) established its rigorous certification process in 1996. It requires a minimum PhD and five years of case work to be a diplomate. The certification exam lasts twelve hours (8:00 a.m. to 8:00 p.m.), and recertification is required every few years. Anderson is one of just 20 experts worldwide who are board certified.

How much of this, I wondered out loud, is about knowing your insects?

"All of it. It's knowing your insects, understanding the field (being well-read), and understanding cases. But it boils down to understanding your entomology to a very high level. ABFE serves to assure a court of a minimum level of education and competence in an expert. ABFE has a strict code of ethics, which means a person could have their certification revoked if they acted in an unethical manner."

Gail volunteers her services for the Innocence Project (IP), whose mission is to redress miscarriages of justice by providing pro-bono legal services to wrongly convicted prisoners, mainly through the presentation of DNA evidence. Anyone who thinks that criminal prosecution is a precise and reliable process should know that since IP's founding in 1992, over 250 exonerations have been won.

One American IP case Anderson was involved in lasted over nine years. Kirstin Blaise Lobato, an 18-year-old, was wrongly accused in 2001 of the sexual assault and murder of a homeless man in Las Vegas. Three forensic entomologists testified in October 2017 that the original case report noted a complete lack of blowfly eggs or larvae on the victim's body at the time it was discovered. Evidently, he had died hours later than originally thought, during which time Lobato was known to be 120 miles away in Panaca, Nevada, where she lived. Such is the reliability with which carrion flies show up on an accessible corpse that their absence is a vital clue to estimating PMI. Lobato was freed in January 2018 after spending nearly 16 years in prison.

"Does this work require a strong stomach?" I asked Anderson.

"Yes, to a certain extent. I'm actually quite a squeamish person. I don't like to see blood and guts on television. But you have to be able to deal with decomposition. You have to be able to confront a dead body and the ooze and the smell. It's kind of gross, and rather upsetting too, obviously, especially if it involves a child. You certainly get used to the smells."

With the ceaseless emergence of new technologies, might

forensic entomology be at risk of becoming old-school? I put the question to Anderson.

"Oh no! There's a lot of interest now in the biology of death: the necrobiome. A hot area is the whole microbe-insect relationship and how that affects the necrobiome."

Convictions and Exonerations

Most of the celebrated cases in which flies have facilitated the solving of a murder have dwelt on convictions, but there are also many exonerations, often less well known, as in the case of a Hungarian ferry skipper. Said skipper was convicted of the knife murder of a man whose body was found on the boat operated by the skipper. The accused was believed to have committed the murder some hours after he boarded the boat at 6:00 one September evening. The presence of fly eggs and larvae from the original autopsy report did not figure in the original trial, only coming to the fore when the case was reopened eight years later. An entomologist testified that the fly species whose larvae were found on the body are not active after dusk. Thus, the victim must have been killed sometime earlier in the day. The falsely accused skipper was exonerated and released.

Dr. Anderson sent me the 2007 report of a legendary case that led to the official exoneration of Steven Truscott, convicted in 1959 of the sexual assault and murder of a friend and classmate, a 12-year-old girl named Lynne Harper, near Clinton, Ontario, Canada. It is a renowned case for several reasons: most notably, it generated such public repugnance toward the sentence of hanging meted out to a minor (Truscott was 14 at the time of Harper's death) that it fueled the demise of the death penalty in Canada.

It also became the subject of a bestselling book by journalist Isabel LeBourdais, who concluded that Truscott's conviction was a miscarriage of justice. After he had been on death row for four months, Truscott's sentence was commuted to life imprisonment in 1960. Paroled in 1974, Steven Truscott maintained his innocence and sought to have his name cleared of the horrible crime. Several decades passed before he got his wish, and evidence from flies proved crucial.

Fortunately for Truscott, Dr. John L. Penistan, the attending pathologist who surveyed the crime scene and conducted an autopsy on Harper's body, also collected and documented insect evidence. In the two days between Harper's disappearance and the discovery of her body in a wooded area known locally as Lawson's Bush, flies of two species had colonized her body. Mr. Elgin Brown, an entomologist, reared the larvae Penistan had collected and was able to identify the blowflies colonizing Harper's face as belonging to the genus *Calliphora*, or bluebottle flies. The flesh flies that had colonized the girl's genital area could only be identified to the family Sarcophagidae. Blowflies, which lay eggs that hatch within hours, are attracted to mucous membranes in the facial area and are not generally seen in large numbers around genitals. Flesh flies, which bear live young, typically colonize the body somewhat later and generally shun the face to avoid competition with blowflies. Being diurnal, neither fly deposits eggs or larvae at night, but the larvae continue feeding.

The science of forensic entomology in 1960 was still rudimentary, and the flies made no appearance in the trial that led to Truscott's conviction. Penistan based his opinion that Truscott was the killer on three circumstantial factors: the state of the contents of the victim's stomach, the degree of decomposition of her body, and the extent to which the body was still affected by rigor mortis. He testified before the jury that he would put Lynne

Harper's time of death as prior to 7:45 p.m. on June 9. Truscott had been seen giving Lynne a ride on his bicycle from their school to a nearby highway at around 7:15 p.m. Truscott's whereabouts were unknown until he returned to the school grounds at 8:00 that evening, at which time he was in the company of others. If Harper's death had occurred at any time after 8:00 p.m., the prosecution's case collapsed.

A reexamination of the insect data nearly 50 years later, based on evidence provided by three forensic entomologists, including Gail Anderson, refuted Penistan's earlier conclusions. The larvae of both species were almost certainly too small to have been birthed before nightfall on June 9. Dr. Sherah VanLaerhoven testified that there was a 95 percent probability that for the blowflies to have reached their meager size of just 2 millimeters at the time of the autopsy, they had to have been deposited in the daylight hours sometime after 11:00 a.m. the following day, June 10. Had they been deposited before sunset the previous night, they would have had the night to feed on the remains and would have grown larger than 2 millimeters.

The entomology evidence raised reasonable doubt that Lynne Harper had died before 8:00 p.m. on June 9. Doubt like that demands acquittal. In 2008, Truscott was awarded $6.5 million for his ten years of incarceration, and for the stigma of living 48 years as a convicted murderer. Truscott's wife, Marlene, got $100,000 in compensation for the time she spent working to clear her husband's name.

To bolster their case, the appellant's team conducted a recreation experiment at the same woodlot where Lynne Harper's body had been found. On June 17, 2006, at approximately the same time and under similar weather conditions, the bodies of three small female pigs were placed in the woods. The unfortunate animals had been electrocuted, and each had a small wound created on the shoulder (with a knife), and a small amount of blood applied to the buttocks and vaginal area, replicating Harper's wounds. These last steps were taken because the presence of body fluids influences the behavior of insects. Within 30 minutes, blowflies of the same genus (*Calliphora*) found on Harper's body had laid eggs on the nose and mouth of each pig. Each pig was observed from the time of placement until dark, and no more eggs were deposited between sunset and sunrise. This experiment rendered it very unlikely that Lynne Harper's body would have remained uncolonized by flies had she been dead before 7:45 p.m. on the day of her disappearance.

Gail Anderson was initially employed by the prosecution, but she ended up arguing for Truscott's exoneration when the evidence was found to support his innocence.*

An Evolving Science

Curiously, despite its promising start in 11th-century China, the field of forensic entomology lay fallow for the next eight centuries. The French army veterinarian Jean-Pierre Mégnin (1828–1905) performed many experiments leading to the recognition of eight distinct waves of insect succession on corpses exposed to air, and two waves on buried bodies. A German contemporary of Megnin's, the physician Hermann Reinhard (1816–1892), focused his attention on buried bodies, and the importance of tiny scuttle flies in being able to colonize them.

*In May 2019, VanLaerhoven and a colleague published a paper in the journal *Forensic Science International* titled "50Years Later, Insect Evidence Overturns Canada's Most Notorious Case—*Regina v. Steven Truscott*," which describes the re-creation experiment (VanLaerhoven & Merritt 2019).

Pekka Nuorteva, formerly with the University of Helsinki, did a lot to advance the field in the 20th century.* He was a major contributor to the first treatise on the subject, *A Manual of Forensic Entomology*, published in 1987. Nuorteva was one of the first to show that insects bioaccumulate toxins, such as mercury, and metals like copper, iron, and zinc. This phenomenon has aided the solving of many challenging cases. For example, when adult flies were reared from the badly decomposed body of an anonymous woman found in the rural area of Inkoo, Finland, they were found to contain unusually low levels of mercury. When she was identified, she proved to have been a student from the University of Turku, located in a geographical region relatively free of mercury pollution.

Subsequently, the detection of drugs like cocaine, heroin, and phenobarbital in maggots removed from human bodies has proved useful in ascertaining deaths from overdose. The fact that flesh fly larvae develop faster on the tissue of corpses containing cocaine and methamphetamine has allowed forensic entomologists to more accurately predict time of death of victims in criminal cases involving drug-related deaths. Heroin also speeds up larval growth but retards development during the pupal stage. When researchers injected various concentrations of morphine into hunks of meat, researchers found higher concentrations of morphine in shed blowfly pupal casings than in adults. Dry insect remains can persist on or around a corpse for a long time, serving as a useful late alternative when suitable tissues are not available.

Identifying insects to species is an important component of forensic entomology, but identifying flies to species can be practically impossible based on physical clues alone. The eggs, larvae,

*I reached out to Nuorteva, but he is now age 94 and his ability to communicate is dwindling.

pupae, and adults of closely related species can be virtually indistinguishable. This is why modern molecular methods—in particular, DNA barcoding—have been invaluable. Although insects change dramatically in their metamorphoses from egg to larva to pupa to adult, their DNA barcode remains unchanged.

The victim's DNA can also be a critical clue for detecting foul play. Happily, DNA survives fly digestion, so molecular analysis of maggots recovered at a crime scene can yield the identity of a body that is no longer present. Gail Anderson illustrated this for me with a case in which a man had moved his murdered wife's decomposing body from the basement of their home when he suspected the police had been tipped off by a suspicious neighbor. When the cops showed up at the house and questioned the man about a smelly stain with a few maggots crawling around on the basement carpet, the husband claimed that the family cat had died there. Laboratory analysis of these maggots turned up human DNA, which put quite a damper on the husband's explanation.

While forensic entomology remains a relatively small field, it is an actively advancing one. The number of practitioners has doubled in the past decade. As yet, there are no university degree programs or dedicated scholarly journals, but there are more than a dozen textbooks, many courses, and at least seven universities offering minors or concentrations in the field. The Third International Meeting of the European Association for Forensic Entomology, held in Budapest on May 25–28, 2016, included such topics as how blowfly larvae could be used to detect the presence of semen at a crime scene, using larval development to calibrate the presence of a synthetic cannabinoid or various concentrations of alcohol, and iFly, a mobile app for on-site forensic data collection. Several new molecular techniques show promise in streamlining the process of identifying fly species and/or victims at a crime scene, including flow cytometry, and next-generation sequencing technologies, such as pyrosequencing, a DNA sequencing method that relies on detecting light emitted by the release of pyrophosphate molecules.

These and other advances since the latter 20th century have made forensic entomology a recognized subdiscipline of both forensics and entomology. It is being embraced by judicial systems worldwide. Today there are dedicated forensic entomology organizations in North America and Europe. A 2009 worldwide survey generated 70 responses from 24 nations.

As the interactions between specific flies and specific bacteria become better understood through detailed studies, a new avenue opens up for the advancement of forensic entomology. We now know that some flies inoculate their eggs, larvae, or both, with specific bacteria, which in turn may be mechanically transferred between the insect and its food source. Presence or absence of a specific bacterial species could provide strong indications as to which flies were present on a corpse and when they left. On the other hand, experts caution that the strong influence that timing and type of microbial colonization may be having on insect development could generate misinterpretations of the period of insect colonization or incorrect estimates of postmortem interval.

Fly activity is not exclusively helpful in resolving violent crimes and other mysterious deaths. It can also be a hindrance. Blood spatters offer important clues to what happened and how, and analysis and interpretation of these clues is a rigorous science. By walking through, imbibing, regurgitating, and/or defecating blood, flies quite literally mess things up. There are three main ways by which flies interfere with blood clues: (1) by walking through blood, flies may change the shape of a blood spatter, which has implications for determining the direction and angle with which the blood has hit a surface; (2) flies transfer blood to other locations; and (3) flies deposit artifacts that resemble blood spatter, a particularly vexing problem because their regurgitate and feces are often virtually indistinguishable from the original blood. Experts are working to develop a comprehensive picture of how flies distort blood evidence.

Not surprisingly, then, forensic entomology is not always viewed as an exact science. Steve Marshall commented to me that while the field has led to the resolution of some murders, it has also contributed to the obfuscation of others. After recounting a few examples of how misidentifications or misinterpretations have led investigators astray, Marshall pointed out some potential sources of error in trying to estimate time (or place) of death from insect evidence. These include geographic location, habitat, season, weather patterns, temperature fluctuations, time of exposure to sunlight (or to artificial light, which may be constant), and a wide range of situations that affect access to the body, which may be indoors or sealed in a car, an appliance, a trash can, or even a garment bag. A hanging body presents no barrier to corpse detection, but it does pose unique challenges to maggots from gravity and the lack of a clear dispersal route when it is time to pupate. Burned bodies tend to be colonized sooner. Then we must consider that each species that feeds on the corpse affects that food resource for the species that arrive later. If certain key species of blowflies are excluded, then the demographics of latecomers will be proportionally affected. These and other factors (for example, perfume, bug repellent, sunscreen, alcohol intoxication) can interact in myriad ways, increasing the importance of case documentation and extensive research to build up a rich library of scenarios and outcomes.

Maggots and Medicine

In addition to murder investigations, maggots have found their way into medicine. Given their associations with filth and decay, the last place many of us would want to have maggots is on infected wounds, yet that is just where we can benefit from having them.

We've known about maggots' ability to facilitate healing for centuries, probably millennia. There are reports of the successful use of maggots for wound healing by Mayan Indians and by aboriginal tribes in Australia. During wars from the Renaissance onward, military physicians noted better wound healing and reduced morbidity and mortality among soldiers whose wounds had been attended by maggots. Napoleon's surgeon general, Baron Dominique Larrey, who reported during France's campaign in Egypt and Syria, in 1798–1801, saw that certain species of fly destroyed only dead tissue and had a positive effect on wound healing.

America's bloodiest conflict provided more fodder (literally) for this unconventional medical assistant. US Civil War casualties, many of whose wounds went unattended for days during triage operations, unwittingly benefited from the presence of flies, which laid their eggs on the open flesh. The soon-hatched maggots fed painlessly on dead and infected tissue without damaging the still-healthy tissue. Gangrene-causing bacteria became maggot food. Not only did maggots devour harmful bacteria, but they converted their food into an additional benefit; substances excreted by the maggots were later found to accelerate healing and prevent the need for many amputations.

These benefits were felt again in World Wars I and II, when similar circumstances once again brought maggots and men together. While tending to soldiers wounded in World War I, an orthopedic surgeon named William S. Baer recognized the efficacy of maggot colonization for healing wounds. He observed one soldier left for several days on the battlefield, a man who had sustained compound fractures of the femur and large flesh wounds of the abdomen and scrotum. When the soldier was delivered to the hospital, he had no signs of fever despite his serious injuries and his prolonged exposure to the elements without food or water. On removing his clothes, Baer discovered that "thousands and thousands of maggots filled the entire wounded area." Yet, to his surprise, when these maggots were removed, "there was practically no bare bone to be seen and the internal structure of the wounded bone as well as the surrounding parts was entirely covered with the most beautiful pink tissue that one could imagine." At that time, the death rate for compound fractures of the femur was about 75 to 80 percent.

More than a decade later, while at Johns Hopkins University, Dr. Baer conducted one of the first scientific studies of maggot therapy. He introduced blowfly maggots into 21 patients with persistent bone infections that had resisted other treatments. Baer observed rapid elimination of dead, festering tissue; declines in pathogenic organisms; reduced odor levels; stabilization of wound beds; and optimal rates of healing. All 21 patients' open lesions were completely healed, and they were released from the hospital after two months of maggot therapy.

This work was published in 1931, the year of Baer's death, and soon thousands of surgeons were using Baer's maggot treatment. Over 90 percent were pleased with their results. A pharmaceutical company, Lederle Laboratories, commercially produced "surgical maggots" until the 1940s for hospitals lacking facilities for breeding maggots.

By the mid-1940s, the antibiotics revolution was under way. These wonder drugs not only cleared up the sorts of intractable lesions that until then only maggots had successfully treated, but they also prevented wounds from becoming infected in the first place. Despite occasional cases of successful maggot-assisted wound healing, the insects fell out of favor.

But nature is crafty and resilient. As agile microbes developed resistance to the drugs developed to suppress them, agile physicians sought alternatives, and maggots began a resurgence. Despite a continued reliance on antibiotics—or maybe because of it—maggots have retained a place in the medical toolbox. Germ-free blowfly maggots are reared in labs, and doctors use the maggots to treat infections and remove dead tissue on patients too weak to tolerate surgery who suffer severe bedsores, traumatic wounds, nonhealing surgical wounds, diabetic foot ulcers, burns, bone infections, and tumors.

The end of the 20th century saw the first controlled, comparative clinical trials of maggot therapy, leading to its acceptance in 2004 as a US Food and Drug Administration–approved method. Dozens of studies have been published since, showing time and again that maggot therapy works better than more conventional treatments. As just one example, a 2012 study from the *Archives of Dermatology* showed that maggots cleared more dead tissue from surgical incisions than did surgical debridement, the often lengthy and painful process by which doctors use a scalpel or scissors to remove damaged tissue or foreign objects from a wound.

Modern-day maggot debridement therapy (MDT) involves the application of sterile (bacteria-free, not neutered) maggots—usually greenbottle flies (*Lucilia sericata*)—to wounds. Collateral damage to healthy tissue is one of the banes of conventional wound treatment therapies like enzyme application, mechanical debridement, or, heaven forbid, surgery. Maggots have a taste for only decaying tissue, not healthy stuff, so they know when to stop. They debride wound sites by removing and dissolving infected and dead tissue, they disinfect by ingesting bacteria, and their wriggling movements help stimulate circulation and promote bruise healing.

The maggots do two types of debridement: mechanical and enzymatic. The first type is produced by movement of the maggots, whose mouth hooks (used to pull their bodies forward) and numerous tiny spines loosen debris with greater fidelity than a surgeon's rasper. Enzymatic debridement is produced by the maggots' digestive enzymes, which liquefy infected and dead tissue into a nutrient-rich brew that can be imbibed by the maggots. Each maggot can remove 25 grams of dead or infected tissue every 24 hours. That converts to a pound of removal a day for every 18 maggots. Maggots do more than remove infected and dead tissue; they secrete allantoin, a compound with antiseptic properties that accelerate the breakdown of dead tissue and promote new cell growth.

Greenbottle fly maggots also release ammonia, which is associated with zealous cleaning by humans, and thus the horrible odor of decaying flesh can be suppressed by their presence.

I looked up Medical Maggots (MM), operated by Monarch Labs, in Irvine, California. MM is America's leading manufacturer and distributor of disinfected fly larvae for the treatment of ulcerative or traumatic wounds. MM's catalog offers products in ten categories, including several types of wound dressings or "cages" that confine the maggots to the wound and prevent them from wandering off to complete their life cycles. A vial of 350 squeaky-clean larvae goes for $250, plus shipping. (The fee is underwritten by donations and waived for patients without insurance or ability to pay.) Typically, 5 to 10 maggots are applied per square centimeter (0.16 square inch) of wound surface area. The wound is dressed, taking care to maintain a degree of porousness so that the maggots do not suffocate, and the maggot application is left in place for 48 to 72 hours.

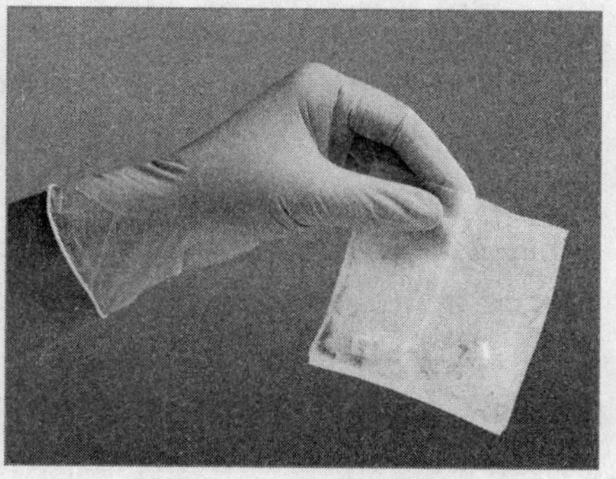

A batch of sterilized young blowfly maggots are sealed inside a fine mesh BioBag, through which they will be able to eat away infected tissue and effectively clean a patient's wound.

(COURTESY BIOMONDE UK)

MM larvae are from the greenbottle fly. These are the same flies I saw carpeting a fresh pile of dog poo in Florida. But fear not, these grubs have been kept in culture for 22 years and are disinfected. Patient anxiety (the "yuck factor") is more often discussed than encountered; rather, patients with persistently infected wounds are only too happy to accept a therapy that can bring relief.

I emailed Dr. Ronald A. Sherman, MM's director and a leader in the field of maggot therapy.

"Would it be fair to say that medical maggots have spared patients from having amputations in the modern era? Death perhaps?"

"The answer regarding amputations is a resounding 'YES,'" Sherman replied. "Published studies of patients whose wounds failed to respond to all conventional care and [who] were scheduled for amputation reveal that 40–70 percent of those patients [treated with maggots] either healed their wounds and avoided amputation, or at least improved the wounds significantly enough that they required much less aggressive surgery.

"Regarding prevention of deaths, my answer is [a] qualified yes. It's impossible to quantify the number of deaths that were avoided. We know that many people die shortly after amputation, and we believe some of those people die because they are no longer as mentally positive or physically active. But we don't know how many fit into those categories, or how many are dying early as a result of the underlying disease (diabetes, circulatory decline, etc.) that led to their amputation in the first place.

"We also know that gangrene often harbors microbes that may appear stable but could spread into the bloodstream and infiltrate the body at any time, resulting in sepsis and death. But quantifying 'what might have happened' with maggot debridement [versus] slower non-surgical removal of the gangrene is pure speculation. Therefore, I can tell you that patients claim that maggot therapy saved their lives, I can tell you that we know it saves lives, and I can tell you there is solid evidence that it improved the quality of lives; but we can neither quantify nor prove scientifically that it saved lives."

"How far do you ship?" I asked.

"I used to ship nearly worldwide, because there was no other source for medicinal maggots. I published my methods for producing the maggots in 1996, and now there are labs around the world, so I can refer people to the most 'local' lab. Routinely, I limit shipping to North America, though I have shipped the insects to non-continental states and territories of the US in the recent past, as well as to countries in Europe, Asia, the Middle East, and South Africa."

"Are there logistical challenges with shipping live insects long distances?" (In hindsight I realize that this was a rather dumb question.)

"Yes. The biggest challenge is keeping them alive."

I was also curious to know if there are any signs that the use of medical maggots is expanding to treating other medical conditions.

"If you mean for wound care purposes for which they are not commonly used and FDA-cleared, then yes. As for medical conditions other than wound care, I don't know. More research is necessary. The maggots have chemicals and activities that also kill germs and stimulate tissue to grow. I find that absolutely fascinating, but I ran out of money to study it. When we better understand those mechanisms, then the maggots (or more likely the maggots' biochemistry) will be available to treat a whole host of additional maladies besides just 'wounds.'"

Researchers are now genetically manipulating these flies to create strains that can deliver a variety of growth factors and antimicrobial agents to hasten healing and tissue regeneration. In addition to some published articles, there are dozens of online videos documenting the effectiveness of this approach.

Like the role of flies in forensics, this is not a weird therapy lying at the margins of medicine; over 2,000 medical centers in the United States have used it. In 1995, medical maggots were being produced in the United States, Israel, and the United Kingdom. By 2002, more than a dozen labs were producing them. As of 2013, some 80,000 patients were being treated by thousands of physicians using maggots produced in at least 24 laboratories and shipped to patients in over 30 countries. Maggot therapy is equally effective for veterinary applications and is regularly used on animal patients.

MDT is also far less expensive than conventional wound healing therapies, and that's nothing to scoff at: the annual cost to treat diabetic foot ulcers in the US alone was $9 to $13 billion in 2013. The revival of maggot therapy has been accompanied by scientific studies that support its effectiveness. Medical Maggots' website includes a list of 69 published studies documenting such benefits as fewer medical and veterinary amputations, more antibiotic-free days, and the successful treatment of burns and other intractable wounds.

Given flies' culpability in transmitting filth and disease, it helps to know that they have atoned somewhat through their contributions to crime solving and wound healing. Recent trends predict that forensic entomology will be with us for the foreseeable future. The future of maggot therapy for wound healing is, perhaps, less assured, but these insect medics remind us that newer techniques are not always better.

I find it remarkable that in a world in which technology advances so quickly, we nonetheless find ourselves turning to maggots to help solve our problems. There is a lesson in that. We are inescapably natural entities. The same fundamental life processes that animate a fly also sustain our own bodies. As we inevitably die and decay, we may become fly food, and occasionally that process may find its way into a courtroom. In another context, becoming fly food may help us regain health and extend our lives.

If flies can do that for us, might we be able to extend compassion toward them?

Chapter 12

Caring about Flies

> It is apparent that the myth of our separation from nature has found a disturbing foothold in our interactions with the members of this vast Lilliputian world.
>
> —JOANNE LAUCK HOBBS

As the dominant and most easily observed group of animals on Earth, insects have always piqued my interest. It was only a matter of time before I committed myself to writing a book about them. But I soon realized that insects in general were too broad a topic for one book—they merit an entire encyclopedia—so I opted for a subset. Flies seemed perfect: diverse, enigmatic, charismatic (if we pause to look more closely), immensely successful, but mostly overlooked. During the three years I cultivated this book, flies unfailingly rewarded my quest to learn more about them.

If you've read this far, then you will most likely have concluded that I'm an advocate of flies. To be so is merely an extension of who I am. Since childhood I have adored animals and despised cruel treatment of them. This was not a reasoned conclusion I'd arrived at but a deeply felt sensitivity long predating my ability to reflect ethically. No creature was vile or loathsome to my sensibilities, and any kid who squashed crickets or stomped ants was far more alien to me than the little beings being mashed beneath their shoes. Whip scorpions and toads inspire no less admiration in me than elephants and sharks. It was no coincidence that after completing six years of graduate school studying bat communication I spent 25 years working for several animal-protection organizations, tackling such issues as killing wildlife to supply school dissection exercises, impoverished housing of rodents in laboratories, and the use of wire snares.

An Entomology Conference

I know that my soft spot for flies is not widely shared by the general public. It isn't even shared by some who study insects as a profession. I was rather surprised by the antipathy toward flies I encountered at a major entomology conference in late 2018. On a shuttle bus from my hotel to the conference venue, the driver described having to go to Las Vegas to collect a coach that had been in repair for two months. The cabin of the bus had been closed, but hundreds of flies had somehow found a way in.

"That bus stank!" said the driver. "Dead flies everywhere."

The implication was that the flies were causing the smell, and there were utterances of commiseration from the dozen or so conferees seated near the front of the coach. This struck me as presumptuous. I've never noticed a bad smell accompanying dead flies on windowsills.

"Couldn't the stink have been produced by whatever it was that attracted the flies?" I ventured.

After a moment of silence, the driver mentioned that a partially eaten apple had been left in the bus. Perhaps this is what drew the flies and created the odor. Or maybe there was another, hidden source of food. Two months is enough time for houseflies to complete their life cycle, and I wonder if perhaps a single female fly, trapped when the bus was originally impounded, had spawned all the flies found dead in the bus.

A few moments later, flies took another hit. Someone mentioned having had an infestation of drain flies. Once again, this brought forth moans of malcontent. "I hate drain flies!" said one passenger. I couldn't imagine why. Perhaps you've seen one or two of these very small, cute bugs—scarcely bigger than this A—whose gray wings make a neat triangle over their backs. Also called moth flies for their resemblance to a tiny moth, they remind me of a stealth bomber in miniature—but unlike a bomber, these flies carry no ominous payload as they rest next to a showerhead or on a tiled bathroom wall, having emerged from a nearby drain, where they graze on accumulated scum and somehow successfully contend with hot water, soaps, cleaners and other chemicals. That doesn't sound like an appetizing diet, but at least they help keep drains clean; one genus, called *Clogmia*, might more charitably be renamed *Unclogmia*. These flies are known to be completely innocuous. I'm happy to see them.

Academic research with flies is often not animal-friendly, as you will have noticed from many of the studies I have described. With the exception of fruit flies for genetics, most research on flies is dedicated to trying to control them either as disease vectors or crop pests, so you can appreciate that a lot of fly research is lethal. "I kill flies!" a research entomologist announced proudly at a fly symposium. Indeed, practically all of the entomologists I met while developing this book kill large numbers of flies as part of their research. But unlike the passengers on the coach, I never detected disdain or indifference among those who specialize in flies. To the contrary, the dipterists I met admire, even revere, the objects of their studies.

Science today is showing unprecedented interest in insects as subjects of ethical consideration. At the same entomology conference—a gathering of over 3,800 insect scientists—I attended a symposium titled "Ethics in Entomology." It was, as far as anyone was aware, a first for this more than century-old meeting. Philosophers, ethicists, and entomologists spoke about eating insects, the question of insect pain, and the wasteful problem of bycatch from field studies that lethally trap and discard masses of nontarget insects. In a paper titled "Why It Is (at Least a Small) Wrong to Harm a Fly," philosopher Jeffrey Lockwood from the University of Wyoming argued that, at the very least, our interactions with insects can be a means of practicing the virtues of mercy, kindness, compassion, gentleness, and love.

This growing ethical concern for insects is starting to show up in published scientific articles. Lockwood's argument is formalized in a 2017 book chapter. A 2015 paper on the role of acoustic communication in the mating behavior of frog-biting midges includes the following ethical note: "Experiments were conducted in accordance with animal welfare guidelines. . . . We encountered no problems using carbon dioxide as an anesthetic, and frog-biting midges effectively recovered from it. Handling and harm to midges was minimized when tethering them by performing the procedure while they were anaesthetized. After the experiments, tethered individuals were euthanized under low temperatures."

Tethering (to the blunted end of an insect pin using superglue) and termination don't sound midge-friendly, but it is notable that methods were judiciously chosen in an effort to reduce harm and potential suffering during the procedures.

Curious to find out if some entomologists are conflicted over having to kill insects to study them, I asked Art Borkent if he has qualms about killing many thousands of little animals:

"Wow. That's only the second time in my life that somebody's asked me that. To me there is a whole aspect of purposely killing something that I've kept to myself all these years. There's a little part of me that is . . . what's the right word? . . . *regretful* isn't the right word. I'm conscious that I'm taking life. There's something different [between] squashing a mosquito on my arm and aspirating no-see-ums off the inside of a net. I see beauty, and to study it I need to retain it. But I'm very conscious of the fact that what I'm trying to study is life. I never go out into the field without having a sense of 'I wish there was another way.'" As we'll see, Borkent is not alone in expressing feelings for a fly.

Research is challenging the old presumption that insects can feel no pain. As this book was nearing completion, a new study, published by a team of geneticists at the University of Sydney, reported that fruit flies suffer a lasting painlike state following injury. Flies subjected to peripheral nerve injury by amputation of one of their legs developed long-lasting hypersensitivity to stimuli not perceived as painful by uninjured flies. Both normal and injured flies tried to escape from a hot plate above 42°C (108°F); however, only injured flies fled cooler temperatures of 38°C (100°F). This sensitivity began five days after the injury and was still present three weeks later. Flies' sensitivity to stimuli normally not perceived as painful, known as *allodynia*, mirrors that of humans and other vertebrates suffering constant pain.

We are still left with the vexing conundrum that we cannot know how flies experience pain because we cannot inhabit their bodies and feel what they feel, but such results should give us pause. Whatever standing on a hot surface actually feels like to a fly, the fact that the fly tries to get off implies that it doesn't feel good. Furthermore, if the little creature can behaviorally avoid (dare I say "dislike"?) something, it follows that it can behaviorally favor ("like") something. In a fly's world, that could be sipping nectar from a flower, basking in the sun, or finding fresh dung. Some philosophers would conclude that flies therefore have intrinsic value. They have interests.

Ecological Anchors

We may choose not to uphold those interests. After all, flies routinely disregard our own, as when they hound us, bite us, and unwittingly infect us with pathogens. But however we may choose to treat individual flies, we would do well to regard them collectively as indispensable components of the world we share with them. Gandhi put it succinctly: "The only way to live is to let others live."

Consider maggots. Their benefits to us are as profound as they are hidden. They are regarded as the most important of all insect larvae for their ability to break down and redistribute organic matter. Without insects, microscopic creatures too small to be profitably eaten by vertebrates would not enter the food chain. By consuming microorganisms, insects bridge the size gap, converting these nutrients into available food for fishes, birds, reptiles, amphibians, and insect-eating mammals as large as bears. Waste products excreted by the larvae provide nutrients at the ground floor of food webs: plants and fungi. Further up the chain of consumption, the bodies of larvae, pupae, and many adult flies are an important food source for larger animals.

Consider also midges. Midges own gold medals for most numerous insects at a given site, and they're consumed by more different species than any other aquatic insect. In their aquatic larval stage, midges are a vital source of fish food. As winged adults, they are no less important to birds. Billions end up in the gullets of shorebirds, swallows, and wrens. While they are among the least charismatic of flies, they are perhaps the most evolutionarily successful and ecologically important aquatic insects on the planet. A recent survey of midges in Canada found a level of diversity that, when extrapolated to the global ecosystem, predicts greater diversity than for all other groups of animals, including the celebrated beetles.

I saw the importance of flying midges to birds firsthand while cycling along the paved footpath bordering Lake Ontario's Bay of Quinte on a late April morning in 2019. Despite nighttime temperatures that were still dipping to near freezing, I had been encountering clouds of midges there since the previous week. Their small black bodies peppered my white rain jacket whenever I passed through one of these clouds. On this morning, equally impressive aggregations of swallows had arrived. I saw at least a thousand along the three-quarter-mile stretch of bay I pedaled. They swooped, circled, and stalled inches above the waterline. Swallows are obligate insectivores, and these weren't eating bees or wasps or beetles or moths, none of which emerge from water. And I'm quite sure I would have been able to detect the larger bodies of aquatic mayflies or stone flies (neither of which are true flies). It was midges that drew their attention. Hordes of tiny flies were nourishing the hungry birds on their northward migration. The swallows' arrival days after the midges' emergence was no coincidence; it's been going on for thousands, perhaps millions, of years.

I wonder, though, whether we, unlike the swallows, are losing our connection with insects. It's a question that interests a growing number of scholars. As humankind becomes increasingly urbanized around the globe, do we risk becoming more and more alienated from nature and the myriad ways it benefits us? American journalist Richard Louv thinks so. In his influential 2005 book *Last Child in the Woods*, Louv introduced the idea of *nature-deficit disorder*, in reference to the possible negative consequences to individual health and the social fabric as children lead increasingly urbanized, indoor lives away from physical contact with the natural world. A few years earlier, American botanists James

Wandersee and Elisabeth Schussler coined the term *plant blindness* for the lost connection between the foods we eat and the crops that provide them, and the missing awareness that we depend on plants for survival. I propose *insect blindness* for our failure to recognize their indispensable role in sustaining us by their services as pollinators, food web components, pest controllers, and janitors. While we're at it, how about *ocean blindness*, since most of humanity is estranged from that habitat, which provides more than half the world's oxygen. And since the ocean would not function without fish life, and vice versa, we may add *fish blindness*.

You get the point. It's interdependence. To paraphrase John Muir: "When one tugs at a single thing in nature, one finds it attached to the rest of the world." Our planet functions as an interactive whole. Start removing or damaging components of this whole, and deterioration ensues. Continue mucking things up, and sooner or later the whole system collapses. That's what happened to the Easter Islanders when they cleared their island of all trees; and to the Mayans when overpopulation, environmental destruction, and constant warfare left them ill prepared for drought and food shortages.

In the preface to their 1983 book *Extinction: The Causes and Consequences of the Disappearance of Species*, ecologists Paul and Anne Ehrlich devised a fitting analogy for the perils of biodiversity loss. Consider our planet represented by a jumbo jet. Each of the millions of rivets holding the fuselage together represents an individual species. Losing a species equates to popping a rivet from the jet. Hundreds, perhaps thousands of rivets could be popped out at random, and the jet will continue to function as a whole. But if the process is allowed to continue, pieces of the fuselage will start to loosen and rattle. Inevitably, as this "extinction" process continues, a chunk of the jet will fall off. We know what happens next: collapse. The entire system goes down. Diversity fosters stability. We can run roughshod over the planet for only so long before our behavior catches up with us.

"Bugpocalypse"

And it is catching up with us. Insects are fast disappearing. According to the best available data, the total mass of insects is falling by a precipitous 2.5 percent a year, a rate of loss (and likely extinction) eight times faster than that of mammals, birds, and reptiles.

A study published in autumn 2018 documented a 76 percent decline in the total biomass of flying insects (crawling bugs were not sampled) netted at 63 locations in Germany over the last three decades. Losses in midsummer, at the peak of insect abundance, exceeded 80 percent. Pesticide use and loss of suitable habitat to farming are suspected as the leading culprits. One of the study's coauthors described the implications thus: "If we lose the insects, then everything is going to collapse." *The New York Times*, in a somber editorial, described it as an "insect Armageddon."

The "bugpocalypse" appears to be a global phenomenon. In 2014, an international team of biologists estimated that the numbers of invertebrates around the world has dropped by nearly half since 1980. In a pristine Puerto Rican rain forest, depending on the sampling method used, invertebrate abundance was between 4 times and 60 times lower in 2012 than it had been in 1976. During this time, mean maximum temperatures there rose by 2 degrees Celsius. David Wagner, an expert in invertebrate conservation at the University of Connecticut, referred to it as "one of the most disturbing articles I have ever read."

It isn't known how many fly species have joined the growing

list of species extinctions. Considering that most species remain undescribed, an unknown number are being lost before we even know they exist.

Observant citizens are noticing the declines. A French translator shared this with me: "My husband and I have often remarked, after a longish drive, that the windscreen was virtually clear of insects now, whereas in the old days one had to stop every couple of hours to clean the splatters of blood and various yellowish matters, so dense they impaired the driver's view. What happened to all those flies???"

The impact of automobiles themselves is not trivial. A six-week-long survey of butterfly roadkill in central Illinois tallied more than 1,800 dead butterflies. Extrapolated to all of Illinois, this suggests that 20 million butterflies die on roads there every week. All else being equal throughout the fifty US states, that would translate to about 1.3 billion butterflies succumbing to drivers over a three-month span of summer weather. Densities of flies, beetles, bees, and wasps are usually higher than butterflies, so presumably their casualty rates are proportional.

Art Borkent's perspective, that of a professional entomologist, is not much different from the French translator's: "I don't know anybody who does what I do—going out and collecting species, killing them, and describing them in minute detail—who doesn't have a profound awareness of the sense of loss, the extinctions we're witnessing right before our eyes today. I've been privy to conversations among dipterists over the years and there's this collective sense of how we're screwed, that we're losing something very precious and beautiful, and that we're in deep trouble."

With their abundance and diversity and their vital contributions to healthy, functioning ecosystems, it follows that insect declines reverberate in populations of other creatures in the same ecosystem. So it goes with the insectivorous lizards, birds, and frogs that are also slipping away in the Puerto Rican study mentioned above. Looking north, the total wild bird population in North America has dropped by almost a third, or about 3 billion individuals, since 1970. This decline encompasses a wide range of species and habitats, not just endangered species but also common backyard birds. So it goes also with marine life, half of which we have lost since 1970, and if you've studied the history of commercial fishing, then you know that we had already lost a great deal before then. No wonder American philosopher Jeff Lockwood observes, "If absence makes the heart grow fonder, humans should be head-over-heels in love with nature." I've seen no statistic more telling of just how thoroughly we are ensconced in the Anthropocene Epoch as this one: of the entire biomass of terrestrial vertebrates on Earth now, wild animals make up only about 3 percent of the total, whereas humans comprise a quarter, and livestock the remaining three quarters. The ratios are essentially unchanged if we take just mammals (no fishes, birds, reptiles, or amphibians): 60 percent livestock, 36 percent humans, and all the rest—elephants, hippos, whales and dolphins, giraffes, rodents, bats, monkeys, etc.—just 4 percent! Our ponderous footprint isn't just human-shaped; it is a pig's or a cow's or a goat's hoofprint, or the three toes of a chicken or turkey—animals we raise in astronomical numbers to kill and eat.

It's impossible to attribute such a profound reshuffling of life on Earth to any one cause, but the so-called Sixth Great Extinction is of our own making. The overwhelming and still growing human presence generates a multitude of threats to nature: urban encroachment and habitat destruction; air and water pollution; agricultural intensification, especially animal agriculture; commercial fishing and aquaculture; hunting and poaching; and the long present but only recently broadly acknowledged climate emergency.

Fly Friends

As a biologist who earns a living mainly by writing and speaking about animals and their remarkable abilities, I consider animals my clients and my friends, and like any shrewd collaborator I try to avoid harming or killing them. There are exceptions. I have terminated ticks after discovering them sunk into my skin, and I have had Lyme disease. I have treated myself and my now grown-up child for head lice, and I have combed out and killed fleas from infested cats. I have also killed a significant number of biting flies, mostly mosquitoes in the act of trying to scalp me. I've slapped blackflies and no-see-ums. Once, plagued by deerflies on a canoe trip, I kept a tally of successful slaps as they alit on my head, tallying over 100 kills. (I have since discovered that a hat is quite an effective barrier against deerflies.) And I have on rare occasions managed to outmaneuver a stable fly nipping at my ankles.

But these are exceptions, not routines. My rule of thumb is that I attempt to eliminate them only in self-defense. If I know they're out for my blood, then they're fair game. Even then I often choose to exercise restraint. Countless times I have spared mosquitoes the slap, and while I may try to catch a persistent horsefly, I am loath to kill one. However nefarious their intentions toward me, I try to retain a reverence for their completeness and their rightful place in the web of life.

I'm not alone. Perhaps you are one of a growing number of humans who consider an insect not as a pest or a threat but as a fellow denizen of the planet. It's a viewpoint more common in Eastern religions than Western ones. A sign on an entrance door at a Vipassana meditation center reads: PLEASE TAKE CARE NOT TO KILL ANY INSECTS IN THE DORMITORY. "When I see insects in my home, I don't kill them," says neuroscientist Christof Koch, speaking of the influence on him of the Buddhist teaching that sentience—the capacity to feel—is probably everywhere at varying levels.

An ethic of live-and-let-live seems on the rise in Western culture. The High Park Mothia, a volunteer group of mostly amateur entomologists, have been setting up light traps in Toronto's High Park since 2016, then photographing the catch. So far, they have documented over 900 moth species, including one not seen in the region for over 100 years and believed locally extinct.

The group has a strict no-collection, no-kill policy. Curious to know why, I asked the Mothia's director, Taylor Leedahl, a professional dog walker and owner of TinyHorse, a company that creates gear for managing multiple dogs.

"Most of the people involved have a high regard for the insects we're looking at, and we would never want our investigations to negatively impact them. We are there to witness only. I think people have a more memorable, important experience when they interact with a living organism rather than just annihilating it."

I had to ask Leedahl, "Do you ever see flies on the illuminated sheets you set out?"

"Yeah, absolutely. Ever since we started doing this, we've talked about expanding the work to monitoring insects in general."

When one stops to consider that 900 species of moth can inhabit a 400-acre park in a big city with long, harsh winters, one may appreciate how rich biodiversity can be in urban places, and the importance of having green spaces there. Leedahl laments an apparent decline in numbers of amateur naturalists, but she's proportionately encouraged by the power of citizen science nature apps like iNaturalist to reengage people, especially youth, with nature.

Having a soft spot for moths is one thing, but can we extend feelings of kinship to a fly? Can we engage our deep potential for empathy toward creatures so widely disliked and rejected?

John Pierre can. An author, professional speaker, and fitness trainer to Ellen DeGeneres and other notables, John has a *reverence for life* that would have brought a smile to Albert Schweitzer when he coined the term while watching hippopotamuses from an African riverboat in 1915. When I last met John, he told me, eyes sparkling in their usual way, that he had rescued eight houseflies stranded in his apartment the prior week. That he keeps count shows his commitment. John also uses the tip of a paper towel to mop up tiny phorid flies unable to escape the surface tension of a glass of water. The tiny blot of water soon evaporates, and the rescued fly lives another day. I aim to try it.

John Pierre is not the only friend of flies. A Hawaiian friend of mine told me that her late stepfather would kick back in his recliner holding a can of beer, with flies buzzing around him, and declare, "The flies are my friends!" Another friend told me how saddened he would feel, years ago, when he noticed the twisted strips of flypaper hanging from the ceiling of an Idaho diner, littered with flies, most dead, others struggling in the stickiness. "I suppose," he speculates, "they die of exhaustion and starvation."

Celebrities also are getting behind insects. Champion golfer Rory McIlroy made news when he carefully removed an insect (a camera close-up suggests a beetle) from a putting green during the televised 2019 PGA tournament, depositing it in a safer place nearby before draining a 20-foot putt. "He deserves to make this after that," said the commentator. "That's what you get for being nice to wildlife!" Paul Rudd, star of the 2015 film *Ant-Man*, abstains from killing insects for the simple reason that he doesn't believe he's better than them. Since 2014, actor Morgan Freeman has been speaking out for honeybees. Freeman installed 26 hives on his 124-acre Mississippi ranch, converted it to a sanctuary, and has openly criticized the destruction wreaked by Roundup and other broad-spectrum pesticides. Recording artist Moby removed his swimming pool to make room for garden trees and flowering plants tailored to the needs of bees, calling it a "much better use of a backyard than a dead concrete hole."

If all this beneficence is causing you to shake your head, consider these words from Joanne Lauck Hobbs's insect book *The Voice of the Infinite in the Small*: "What may sputter inside ourselves in indignation at the thought of helping a fly may only be self-importance arising out of a narrow band of awareness. Our sense of self expands when we extend our compassion to insects."

Goodness is not an exhaustible commodity. If you've rescued a ladybug from a glass of water or a cricket from a swimming pool, then you know from experience that even the smallest act of Good Samaritanism feels good. I've watched butterflies and honeybees that I scooped moribund from the road recover their strength and fly away after being fed sugar water. In rescuing a fly, John Pierre nurtures his spirit. It's not why he does it, but it is a benefit nonetheless. If you doubt this, try it.

To those who would sooner reach for a can of Raid than an eyedropper, take note that our widespread aversion to insects is probably more learned than innate. There is evidence for an innate human fear of spiders and snakes, but these are rare exceptions. Flowers, houseflies, and fishes, for instance, trigger no such aversion. "We are born without the fear of nature," writes biologist Piotr Naskrecki in the introduction to his 2005 book *The Smaller Majority*, which showcases mostly insects. "Young children are fascinated with life around them, equally intrigued by a caterpillar or a dog. The fear of most creatures is instilled in us later in life by overly protective parents or teachers, peer pressure, and misguided media. By the age of ten most children either love or hate insects and other tiny organisms."

Within the grand legions of insects on Earth, while the ants stand out as the military establishment, the flies are the entrepreneurs and the con men. Evolutionarily agile, commonly deceptive, and often harmful to those who associate with them, flies are easy to dislike and hard to love. But behind the ranks of the infamous among them—the biters, the vectors, the flesh eaters, and the muck munchers—there is a vast minuscule world of the obscure and the beautiful: delicate long-legged flies who skitter across leaves in their electric-gold cloaks, the gossamer outstretched wings of an amorous fungus-loving fly, the spectacular caliperlike head ornaments of an antler fly, sparring male cactus flies squaring off like aliens on stilts (see the image in the photo insert), or the bullish shape and yellow shag-carpet fuzz blanketing the body of a bumblebee-mimicking flower fly.

We are taught from an early age to shun flies, and I have not been immune to our deeply entrenched cultural aversion to these insects, but as I continued to plumb the known depths of their lives, any aversions receded and my heart softened. As I researched and wrote this book, in cafés, in libraries, and at home, I was visited by dozens of flies. Far more of them shared my workspace than any other type of visible organism. They preened on my laptop, darted across the backlit screen, sipped stray stains from tabletops, and made brazen explorations across my arms and hands. No season or climate was beyond their reach. One tiny client even paid a visit to me in the depths of a Canadian winter, landing on my music score while I sang at a church Christmas service.

"No animal or plant in nature is capable of ugliness unless we disapprove of it," writes novelist and naturalist Jonathan Franzen in his 2018 book *The End of the End of the Earth*. During the almost 60 years since I started toddling around in a backyard gazing at insects, Franzen's sentiment has become apparent to me. By rejecting cultural norms of intolerance, I've come to enjoy the light tickle of a housefly's feet against my skin as it runs about, dabbing, tasting with its padded feet and imbibing through its spongy proboscis.

A small fly preens itself atop the author's computer monitor.

(PHOTO BY THE AUTHOR)

I like the subtle habits of flies. I like the way houseflies make staccato darts across a surface with small, jerking movements so fast they appear to be gliding. I like that I am less likely to feel it when a fly alights on me than when it takes off again—the tiny push of their legs against my skin. I like how a housefly's proboscis descends, usually shortly after landing, pressing and spreading against the surface like the soft foot pads of an elephant. I like knowing there is a fly whose waxy, hairy body armor traps air, allowing it to dive.

I also like the urbaneness of flies. At a coffee shop in downtown Delray Beach, Florida, I noticed three tiny flies on chrysanthemum stems inside a large glass vase. At first I thought with some regret that they were probably doomed to perish on a windowsill or when a night custodian does the rounds. But the flies weren't feeling trapped. They were courting animatedly, waving their wings and darting and feinting on the greenery like spirited dancers.

Insects are integrated into our lives, even our physical composition. "More than a quarter of the world's population eats insects," says journalist David MacNeal, author of *Bugged*. I wish to qualify MacNeal's statement with the word *intentionally*. If we add inadvertent intake, then virtually *all* humans are eating insects every day. Insects' ubiquitous presence in the grains, fruits, and vegetables we consume means that almost anyone who eats ingests dozens of insects or insect fragments every day.* Being part of a food web inevitably means you ingest things you don't mean to. The presence of beetle fragments in our breakfast cereal is as inevitable as the presence of pus cells in the cow's milk one may pour onto it (and one of the reasons I favor plant-based milks).†

How closely, then, are flies' fates enmeshed with our own? Gail Anderson shared with me a blunt perspective uniquely fitting a forensic entomologist: "Without carrion insects we'd be dead. The Earth would have used up its nutrients a long time ago. We are all bags of nutrients, and flies recycle those nutrients back to the Earth. This not only stops disease from lingering, but it provides plants with their food. Life continues."

For all our efforts to suppress them, our large ecological presence has been a boon to many flies—all those fruit orchards, all those livestock, all those dead bodies, all that excrement, all that compost. To be sure, our destruction of wild species has devastated, indeed annihilated, many of the world's more obscure flies. But let's not kid ourselves: a million years after the last human is gone, a fly will be perched on a leaf or rock rubbing her feet

*As just one example, the US Food and Drug Administration (FDA) permits up to 450 insect parts and 9 rodent hairs in every 16-ounce box of spaghetti: https://www.cnn.com/2019/10/04/health/insect-rodent-filth-in-food-wellness/index.html (accessed May 12, 2020).

†According to recent FDA data, there are about 5 million pus cells in the average cup of cow's milk: https://nutritionfacts.org/2011/09/08/how-much-pus-is-there-in-milk (accessed May 12, 2020).

together. We may imagine a world without flies, but should it come to pass, we won't be there to witness it.

And so I close with a single fly and a single human. The aviator Charles Lindbergh, delirious with sleep-deprivation, is alleged to have had conversations with a fly during his historic 1927 transatlantic solo flight. In the movie *The Spirit of St. Louis*, Lindbergh (played by Jimmy Stewart) discovers the fly near the beginning of his 33-hour flight. The fly is credited with averting possible disaster when it alights and walks around on the pilot's cheek, waking him from a doze during which the plane was losing altitude. There is even a filial moment between man and fly: when Lindbergh is over Greenland, he informs the insect that this is the last chance to leave and have land beneath its wings for the next 1,800 miles. Appearing to take the hint, the fly exits through an open window.

When I learned of Lindbergh's encounter with the fly, and knowing that the pilot was destined for glory, my sympathies were aroused for the little insect. The empathy centers of my brain glowed with concern for its uncertain fate. I believe such feelings are a microcosm of how our relationship to flies, indeed all life, can evolve. Whatever justified angst and antipathy we may feel for certain flies, we can simultaneously cultivate respect, even reverence, for their essential place in the world. Failure to do so is not merely a moral mistake; it is a fatal ecological error. Like it or not, our fate is bound to theirs. Lindbergh's safe landing at Paris's Le Bourget Field is a metaphor for our future.

Acknowledgments

A lot of people helped with this project. I gained a wealth of help and support from many scientists, who were always willing to give kindly of their time: Stephen Marshall, Glenn Morris, Art Borkent, Stephen Gaimari, John Wallace, Brock Fenton, Mark Deyrup, Robert Voss, Marla Sokolowski, Kelly Dyer, Bill Streever, Gail Anderson, Bob Armstrong, Tamara Szentivanyi, James Thompson, Ashley Kirk-Spriggs, Eric Benbow, Patrick O'Grady, Dave Taylor, Leo Braack, Christine Johnson (AMNH), Ila France Porcher, Shelly Adamo, Paul Bedell, Terry Whitworth, Mike Howell, Thomas Pape, Henry Disney, Jeff Tomberlin, John Diehl, Taylor Leedahl, Priscilla Tamioso, Jussi Nuorteva, Jeffrey Lockwood, Galit Shohat-Ophir, John Hudson, Norman Woodley, and Martin Hauser.

Special thanks to David Grimaldi at the American Museum of Natural History for sexing Robert Voss's botfly, and to Martin Hauser of the California Department of Food & Agriculture for sexing a horsefly. Also, thank you Stacey Gordon at the Jameson Law Library, University of Montana, and Morag Coyne at the Douglas Library, Queens University, for research help, and to Emily Balcombe for organizing the references.

For photo images, special thanks to Stephen Marshall, who kindly donated several images from his breathtaking book *Flies: The Natural History and Diversity of Diptera*. Thanks also to Brock

Fenton, David Grimaldi, Joseph Moisan-De Serres, Anton Pauw, BioMonde UK, Ronald Sherman, Ximena Bernal, Vinayaraj V R, Romano Gallai, Martin Hauser, John and Kendra Abbott, Vincent Pang, Karolina Stutzman, and Karen Mitchell.

Bob Armstrong, Pavel Volkov, and Sebastian Moro directed me toward several arresting videos and useful studies.

For ideas and personal encouragement, thank you to Susan McCourt, Maureen Balcombe, Joe and Anthea Messersi, Ken Shapiro, Martin Stephens, Patricia Gabaldon, Joanne Lauck Hobbs, Dori Erann, Adriana Aquino-Gerard, John Pierre, Carrie P. Freeman, and Michael W. Fox.

To the editorial team at Penguin Random House, especially my editor, Matt Klise, thank you all for your enthusiasm and for being a joy to work with. To my wise and attentive agent, Stacey Glick, thank you for recognizing from the start the potential of a project that might have prompted questions about my sanity.

To you, the reader of this book, thank you for your curiosity about all the wonders that make whole this beautiful, enigmatic planet.

I take full responsibility for any factual errors, and I invite updates and corrections.

Notes

Chapter 1: God's Favorite

6 **"There are no apologists for flies":** Deyrup 2005, p. 112.
9 **At any one time, there are some ten quintillion:** McGavin 2000.
9 **That's 200 million for every living human:** Grzimek 2003.
9 **1.4 billion insects for every human:** MacNeal 2017.
9 **Ants alone are thought to outweigh:** Grzimek 2003.
9 **termites outweigh us by a similar:** Margonelli 2018, p. 10.
9 **researchers at the Animalist channel:** Farnham 2018.
9 **British fly expert Erica McAlister:** Gorman 2017.
10 **Phil Townsend, a remote-sensing specialist:** "Hotspot for Midges Proves to Be Fertile Ground," *Nature* 454 (August 13, 2008): 815. https://doi.org/10.1038/454815f (accessed June 24, 2019).
10 **Some phantom midges amass:** Marshall 2012.
10 **There are more living organisms:** Zlomislic 2019.
10 **A British biologist estimated:** McNeill 2018.
10 **According to a 1998 estimate:** Whitman et al. 1998.
11 **"If Canada possesses":** Hebert et al. 2016.
16 **If each occupied a ⅛-inch cube:** Teale 1964.
17 **Even Antarctica is home to a few intrepid midges:** Marshall 2012.
17 **Some northern midges can dehydrate themselves:** MacNeal 2017.
17 **Other midge larvae live over 1,000 meters:** Linevich 1963, cited in Armitage et al. 1995.
20 **A 2007 Major League Baseball playoff game:** Davidoff and King 2017.

20 In August 2018, a fly sabotaged: www.dw.com/en/fly-ruins-german-domino-world-record-attempt/a-44955761.

20 In pre-17th-century Western painting: Klein 2007.

20 During the Renaissance: Berenbaum 2003.

21 Over the course of months, the colored blots accumulate: Stinson 2013.

21 You can listen online to Hemsworth: www.youtube.com/watch?v=qjLBXb1kgMo.

21 In a sultry 1999 song titled "Last Night of the World": www.youtube.com/watch?v=02TUsZzF6es.

22 consider what Winston Churchill remarked: Bonham Carter 1965.

22 There isn't much mystery as to why: Howard 1905.

22 Someone was feeling playful: McGavin 2000.

22 Because it featured a bright yellow abdomen: https://www.youtube.com/watch?v=VWYRXP50jBc.

23 other insects are named: www.telegraph.co.uk/news/2017/04/12/organisms-named-famous-people-pictures/.

Chapter 2: How Flies Work

29 Despite notable competition from butterflies: Grzimek 2003.

31 An insect's small size confers: Sverdrup-Thygeson 2019.

31 A housefly attains 345 beats per second: Lauck 1998.

31 and a tiny biting midge, a startling 1,046 beats: Sjöberg 2015.

31 Despite its outlandishly enlarged eyes: Marshall 2012.

33 The gearbox resides at the base: Deora et al. 2015.

33 Special nerve cells detect the twists: Oldroyd 2018.

34 The fly walks by altering the angle: Chinery 2008.

34 This early-warning system: Chinery 2008.

35 "They fly fast": Witze 2018.

36 The insect's compound eye has been: Pomerleau 2015.

37 Our own visual systems produce: Blaj and van Hateren 2004; Kern et al. 2006.

37 As Peter Wohlleben says: Wohlleben 2017, p. 23.

38 In carefully controlled slo-mo filmed: Card and Dickenson 2008.

41 Fly authority Stephen Marshall suspects: Marshall 2012.

41 Today's flies are 100 times as sensitive: Sverdrup-Thygeson 2019.
41 Each pore houses individual neurons: Shanor and Kanwal 2009.
41 The fifth cell doesn't aid taste: Barth 1985.
42 Quite simply, a full fly isn't interested: K. Scott.
43 There is already a device called: www.mosquitomagnet.com/advice/how-it-works.
43 Lower sounds allow more amplification: www.bernstein-network.de/en/news/Forschungsergebnisse-en/fliegenhoeren.
44 Long-term exposure to high decibels: Galluzzo 2013.
45 "It's a great gig because": Guarino 2017.
45 Another threat is that sunscreen: Pennisi 2017.

Chapter 3: Are You Awake? (Evidence for Insect Minds)

49 By constructing an escape route: Heinrich 2003.
49 The authors conclude that: Barron and Klein 2016.
53 Cornell University entomology professor: www.youtube.com/watch?v=1WoS3lG7LUs&feature=youtu.be.
54 A notable case is the detour behavior: Tarsitano and Jackson 1997.
54 A more recent study by the same: Cross and Jackson 2016.
55 In trials in which choosing: Sheehan and Tibbetts 2011.
55 This behavior indicates: Gallup 1970.
56 Brown dots, which matched the ants' body color: Cammaerts and Cammaerts 2015.
56 It reminds me of the famous words: Goodall 1998.
56 This technique enables an ant: Maák et al. 2017.
56 A New World ant of arid desert regions: Möglich and Alpert 1979.
56 Digger wasps use flat pebbles: Brockmann 1985; Griffin 1992.
57 One bug caught and chugged: McMahan 1982, 1983, cited in Pierce 1986.
57 The ants learned to favor: Maák et al. 2017.
57 They can recognize human faces: Dyer et al. 2005.
57 They understand the concepts: Muth 2015.
57 Bees also seem to understand: Howard et al. 2018.
58 "This suggests that the bees": Perry and Barron 2013.

58 And dopamine and serotonin: Van Swinderen and Andretic 2011; Miller et al. 2012.
58 Like ours, fly brains manage: Ofstad et al. 2011.
59 In the fruit fly this ability: Klein and Barron 2016.
59 These memories persist: Reviewed in Giurfa 2013.
59 Another hallmark of attention: Reviewed in Giurfa 2013.
59 but when the researchers tapped: Shanor and Kanwal 2009.
59 Their need for sleep rises: www.uq.edu.au/news/article/2013/04/flies-sleep-just-us.
60 Analysis of their large data set: Arbuthnott et al. 2017.
60 To explore the possibility: Kiderra 2016; Grover et al. 2016.
60 The brains of noncourting flies: Grover et al. 2020.
61 Females who could not observe: Mery et al. 2009.
61 In another experiment: Mery et al. 2009.
61 "I'll have what she's having!": Young 2018.
61 "How can we be certain": Griffin 1981.
62 Even so, they recommend anesthetizing insects: Eisemann et al. 1984.
62 The eminent insect physiologist: Wigglesworth 1980.
62 Insects don't limp on injured limbs: Alupay et al. 2014.
63 In Dawkins's words: Dawkins 1980.
63 The rig is designed so that: Heisenberg et al. 2001.
63 Flies soon learn to keep to: Putz and Heisenberg 2002.
63 Praying mantises, crickets, and honeybees: Sources cited in Groening et al. 2017.
64 It has been known for decades: Colpaert et al. 1980; Danbury et al. 2000.
64 The team tentatively concluded that: Groening et al. 2017.
64 When the training was reversed: Yarali et al. 2008.
64 Fruit flies also exhibit: Tabone and de Belle 2011.
64 That they respond similarly: Dason et al. 2019.
65 The sensory system used with bees: Perry and Barron 2013.
65 Studying feelings in insects: Perry and Barron 2013.
66 This suggests motivation: Krashes et al. 2009.
66 Lastly, when subjected to a stressful situation: Gibson et al. 2015.
66 The authors of this study: Gibson et al. 2015, p. 1403.
67 To test for personality in fruit flies: Kain et al. 2012.
68 The study authors note: Kain et al. 2012.

Chapter 4: Parasites and Predators

73 **"Great flies have little flies":** De Morgan 1872.
74 **In other species, however:** Evans 1985.
74 **These dangling morsels are eaten:** Marshall 2012.
75 **If you think that sounds gross:** Spielman and D'Antonio 2002; Art Borkent, personal communication, July 2019.
76 **Spore release is delayed until sunset:** Zimmer 2000.
77 **And despite their name:** www.sciencedirect.com/topics/medicine-and-dentistry/dermatobia-hominis.
78 **The mammoth botfly is known:** Marshall 2012.
78 **It also happens to be one of Africa's:** Marshall 2012.
84 **Some of the most charismatic:** Porter 1998.
85 **A 1995 study found 127:** Brown and Feener 1995.
85 **While the headless body stumbles around:** Zimmer 2000.
86 **Ant predation by this fly:** Welsh 2012; Wheeler 2012.
87 **Of a sample of 16 decapitators:** Brown et al. 2015.
87 **Parasitism rates were 1,042:** Bragança et al. 2016.
88 **I learned later that scientists:** Feener and Moss 1990.
88 **The minims have been known to kill:** Zimmer 2000.
89 **four minims may be seen riding shotgun:** Chinery 2008.
89 **In their yearlong study:** Feener and Moss 1990.
90 **Some species in the genus *Atta*:** https://en.wikipedia.org/wiki/Leafcutter_ant#Interactions_with_humans.
91 **The legless, wingless adult females:** Marshall 2012.
92 **Far away from the safety of the colony:** Marshall 2012; original source Disney 1994.
92 **They do not attack from behind:** Deyrup 2005.
92 **Because insect blood does not flow:** Paulson and Eaton 2018.
93 **Most robber flies have a beard:** Deyrup 2005.
93 **Robber flies, of course, have enemies:** See photo in Marshall 2012, p. 261/9.
94 **When I followed up with him:** https://en.wikipedia.org/wiki/Schmidt_sting_pain_index.
94 **visible in McKnight's photos:** www.instagram.com/p/BXdd3SjFI4-/.
95 **So estranged is the egg:** Mortimer 2013.
96 **Wasps are evolving chemical virulence:** Lynch et al. 2016.
96 **"Infected fly larvae actively seek out":** Cell Press 2012.

96 Female flies recognize their wasp parasites: Kacsoh et al. 2013.
97 The only other insects currently: Abbott 2014.
97 "The dialect barrier can be alleviated": Zeldovich 2018.

Chapter 5: Blood-Seekers

100 Bonpland, who was in charge: Wulf 2016.
101 There are about 3,568 described: http://mosquito-taxonomic-inventory.info/valid-species-list.
101 some 110 trillion individuals on Earth: Winegard 2019.
101 Luckily, the great majority of them: Spielman and D'Antonio 2002.
101 Some 1.6 million gallons: Byron 2017.
101 *Before you conclude that these zappers: www.thoughtco.com/do-bug-zappers-kill-mosquitoes-1968054 (accessed May 3, 2020).
101 Someone who may equally be called: Hudson et al. 2012.
102 You may be relieved to know: Smith 2008.
102 That works out to a few hundred bites: Heid 2014.
102 It doesn't help them that their flight speed: Spielman and D'Antonio 2002.
102 According to one study, there may be: Waldbauer 2003.
103 Plastic garbage has also greatly expanded: Berenbaum 2018.
103 That smell can last for years: Art Borkent, personal communication, December 18, 2019.
103 It is not unheard-of for a mosquito: Spielman and D'Antonio 2002.
104 According to a 1966 study: Gilbert et al. 1966.
104 The little biters also shunt impurities: Dowling 2019.
104 A recent study of the mosquito *Aedes aegypti*: Vinauger et al. 2018.
104 Research has shown that mosquitoes: Griggs 2018.
105 Both are filter feeders, straining food: Frauca 1968.
105 Males seek nectar and other plant sugars: "Mosquitoes."
105 This is an immunological response: Deyrup 2005.
106 The serious mosquito-borne: Deyrup 2005.
107 Biting midges have the most diverse: Art Borkent, personal communication, November 27, 2018.
107 Of those that stalk vertebrates: Borkent and Dominiak 2020.
108 During a Skype interview, Art Borkent: Clastrier et al. 1994.

108 **When she drops the dried:** Downes 1978.
108 **Margaret Atwood, who spent:** Mead 2017, p. 42.
109 **Nor are blackflies bound to bloodthirsty habits:** Hudson et al. 2012.
109 **Predators lurk here:** Hudson et al. 2012.
109 **Some dance fly grubs:** Hudson et al. 2012.
109 **Emerging blackfly adults that evade:** Marshall 2012.
110 **With just a few dozen specimens:** McKeever 1977.
111 **On one of their first forays:** McKeever and Hartberg 1980.
111 **Apparently, the elusive males:** Art Borkent, personal communication, November 27, 2018.
111 **Once a suitable frog is detected:** McKeever and French 1991; Camp 2006.
112 **The antennae are incredibly sensitive:** Göpfert and Robert 2000, Bernal et al. 2006.
112 **One estimate finds that a small:** Camp 2006.
112 **The frogs also produce a wide:** Borkent 2008.
112 **Other frogs call above 4,000 hertz:** Grafe et al. 2019.
112 **Producing such calls:** Aihara et al. 2016.
113 **When scientists transplanted:** Halfwerk et al. 2019.
113 **My hunch is that midges listened:** de Silva et al. 2015.
113 **The oldest fossil records of frogs:** Shubin and Jenkins 1995; Wake 1997.
113 **the earliest fossil of a frog-biting midge:** Borkent 2008.
114 **These hairs function as carbon dioxide detectors:** Rowley and Cornford 1972.
114 **And since there were no large:** Borkent 1995.
116 **Instead, their saw-tipped bayonet:** Deyrup 2005.
118 **Once encysted under the bat's skin:** Marshall 2012, p. 404.
119 **These flies are found virtually worldwide:** Marshall 2012.
119 **Thus, mosquitoes found at a crime scene:** Curic et al. 2014.
120 **This approach could be used:** Hoffmann et al. 2016.
120 **Human sleeping sickness is transmitted:** Pearce 2000.
120 **So vital is the tsetse considered:** Armstrong and Blackmore 2017.
121 **Another biting midge has kept:** Art Borkent, personal communication, December 18, 2019.
121 **Five mosquitoes, distended bellies:** Marent 2006, pp. 140–41.

Chapter 6: Waste Disposers and Recyclers

123 **Consider that the average American:** Weisberger 2018.

124 **"every living thing":** Marshall 2012, p. 54.

125 **A pair of biologists:** Wu and Sun 2010.

127 **After a half-ton pile of stable manure:** Teale 1964.

130 **That's what a team of Brazilian researchers:** Moretti et al. 2008.

131 **A Venezuelan study found a similar:** Nuñez Rodríguez and Liria 2017.

131 **The maggot mass also generates:** see citations in Thompson et al. 2013.

131 **This probably explains why egg-laden:** Barton-Browne et al. 1969.

131 **The higher temperatures in the feeding mass:** Rivers and Dahlem 2014.

132 **Bob sent me a short video:** www.naturebob.com/northwestern-crows-eating-maggots-salmon-carcass.

133 **What I didn't anticipate:** "Blowflies and dead lizard." https://www.youtube.com/watch?v=bH3eWPvxrN8.

134 **The soldier fly family to which:** Toro et al. 2018.

134 **The use of black soldier larvae:** Bosch et al. 2019.

135 **Cape Town, South Africa:** MacNeal 2017.

135 **notes Victoria Leung, Enterra's VP:** http://enterrafeed.com/why-insects/.

135 ***According to the United Nations:** Bland 2012.

136 **Enterra touts its product as:** http://enterrafeed.com/why-insects/.

136 **The EnviroFlight representative:** Cindy Blevins, personal communication, October 22, 2019.

137 **Perhaps it is some consolation:** Doherty 2018.

137 **At the other end of the livestock:** Sheppard et al. 1994.

137 **The common housefly is under:** Hussein et al. 2017.

138 **In 2010 it was calculated:** Food and Agriculture Organization of the United Nations 2014.

138 **According to Joanne Lauck Hobbs:** Lauck 1998.

139 **It gets rid of physical and chemical:** http://animals.mom.me/flies-rub-hands-6164.html.

139 **Studies of marked flies:** Teale 1964.

140 **One notably unhygienic individual:** Fullaway and Krauss 1945.

140 John Wallace, a biology professor: John Wallace, personal communication, May 24, 2019.
140 a 2014 Orkin Pest Control survey: McClung 2014.
141 While the maggots munch away: Thompson et al. 2013.

Chapter 7: Botanists

143 "Insects . . . are also our worst enemies": Curran 1965, p. 14.
144 Of the 150 described families: Ssymank and Kearns 2009.
144 About 218,000 of the world's: Marlene Zuk, in MacNeal 2017.
144 David MacNeal puts it well: MacNeal 2017.
145 "Most people just notice": Art Borkent, personal communication, December 2018.
145 The conclusion: "Flies widely replace bees": Lefebvre et al. 2014.
145 During observations of flowering plants: Robinson 2011.
146 A 2016 study by European and Canadian scientists: Tiusanen et al. 2016.
146 They attract flies by providing: Luzar and Gottsberger 2001.
146 As they muscle beneath the swollen: www.naturebob.com/rice-root-lilies-and-blow-flies.
146 The reverse is true for the Neotropics: summarized in Inouye et al. 2015.
147 In a European study, 1,762 insects: Knop et al. 2018.
147 Their influence is also felt: Zimmer 2019.
148 So convincing is the mimicry: Deyrup 2005.
148 Issue 57 (October 2016): *Fly Times*, www.nadsdiptera.org/News/FlyTimes/Flyhome.htm.
149 In all, more than 100 cultivated crops: Ssymank at al. 1998, p. 560.
150 The specimen that Mark "Doctor Bugs" Moffett: www.youtube.com/watch?v=rXVU2WPYcR8.
150 Human noses are unable to detect: Young 2007.
152 Thanks to the careful sleuthing: Gardner et al. 2018.
154 There are risks to such specialization: Session and Johnson 2005.
155 Flies are not routinely beneficial: Missagia and Alves 2017.
155 Male flies pollinate the flowers: Dodson 1962.
156 Other flowers are more inclined: McDonald and Van der Walt 1992.
156 One large orchid subtribe: Pridgeon et al. 2005.

156 The genus *Trichosalpinx* comprises: Bogarín et al. 2018.

156 a dark purple, finely fringed lip: Meve and Liede 1994; Vogel 2001.

157 These vibrational movements might also: Bogarín et al. 2018.

158 The meager proteins serve as a signal: Bogarín et al. 2018.

158 Once inside the myrtle plant: Marshall 2012.

159 It's a costly deception: Jürgens et al. 2013; Jürgens and Shuttleworth 2015.

159 These plant mimics combine a suite: Moré et al. 2018.

159 These flowers are thought mainly: Renner 2006; Policha et al. 2016.

159 Males may get what they're looking for: Renner 2006.

159 Even the female deception: Renner 2006.

159 Yet so irresistible are the sights: Stensmyr et al. 2002; Angioy et al. 2004.

160 whose hatchlings will find no suitable food: Bänziger 1996.

160 So refined is the plant's use: Jürgens and Shuttleworth 2015.

160 A research team from Malaysia: Wee et al. 2018.

160 Experiments using model flowers: Du Plessis et al. 2018.

161 evolutionary shift to saprophilous: Moré et al. 2018.

161 Among the credible theories: Jürgens and Shuttleworth 2015.

161 The aptly named fungus gnats: Pape, Bickel, and Meier 2009.

161 As with flies and flowering plants: Lim 1977.

Chapter 8: Lovers

163 "Flies love having sex": Gorman 2017.

164 Courtship in some long-legged flies: Hudson et al. 2012; Marshall 2012.

164 Males sometimes transfer this fluid: Marshall 2012.

165 Some male stilt-legged flies: Marshall 2012.

165 Male mosquitoes, which may mate: Spielman and D'Antonio 2002.

165 Others try to get away with provisioning: Preston-Mafham 1999.

165 In their efforts to lure: Marshall 2012, illus. p. 285/8.

166 Copulation lasts about two hours: Wangberg 2001.

167 Princeton University researchers discovered in 2016: Coen et al. 2016.

167 Aristotle described the sounds of flies: Keller 2007.

168 Mosquitoes' acoustic fidelity is not immune: Spielman and D'Antonio 2002.
168 One entomologist told me: Spielman and D'Antonio 2002.
168 This schedule avoids the frustrating: Frauca 1968.
168 A research team from Sri Lanka: De Silva et al. 2015.
169 Courtship singing may be the source: Runyon and Hurley 2004.
169 But a theory published in 1975: Zahavi 1975.
171 Male *Drosophila* fruit flies will spar: Yurkovic et al. 2006.
171 Here, in roughly escalating order: Zwarts et al. 2012.
171 Such hierarchies require the ability: Yurkovic et al. 2006.
171 Winner/winner, loser/loser: Yurkovic et al. 2006.
172 For up to two minutes, they grapple: Marshall 2012.
172 If he scores, he will immediately: Evolutionary Biology Lab, University of New South Wales. www.bonduriansky.net/waltzingflies.htm.
173 Ken Preston-Mafham, who has studied: Preston-Mafham 2006.
173 Male-mounting could also be a way: MacNeal 2017.
173 Vinegar flies transfer a chemical: Grzimek 2003.
173 Or she may take a slightly less blunt: Sokolowski 2010.
174 Reluctant to attribute emotions: Sokolowski 2010.
174 "The vagina is an elongate muscular tube": Puniamoorthy et al. 2010.
174 Some caddisfly penises: Scudder 1971.
174 One picture-winged fly has a coiled penis: Thornhill and Alcock 1983.
175 an entire book devoted: Wangberg 2001.
175 During an academic-library visit: Theodor 1976.
175 When Nalini Puniamoorthy: Puniamoorthy et al. 2009.
177 The standing record for continuous sex: Pearson 2015.
177 They reached Pensacola by 1949: Stiling 1989.
177 The speeds were 44 and 51 meters per minute: Evans 1985.
178 In a 2010 study, Puniamoorthy and Kotrba: Puniamoorthy et al. 2010.
179 Incidentally, both sexes of scavenger flies: Marshall 2012.
180 The scientists admit that their data: Briceño et al. 2007.
180 The authors conclude: Briceño et al. 2015, p. 403.
181 If you can get past the fact: www.youtube.com/watch?v=ttqU79TsoX8.
181 They tend to become inebriated: Brookes 2001.

182 **But when they placed the flies:** Yong 2018.
183 **"If the reward system is saturated":** Coghlan 2018.
183 **This complements the findings:** Shohat-Ophir et al. 2012.
183 **"It may not be a myth":** Quoted in Brown 2013.
184 **So it seems that both females and males:** Shao et al. 2019.
184 **Reproductive senescence is a well-documented:** Committee Opinion No. 589, 2014.
184 **Reproductive senescence also happens:** Miller et al. 2014.
184 **The dopamine system:** Neckameyer et al. 2000.
185 **A few species, like the tsetse fly:** Rivers and Dahlem 2014.

Chapter 9: Heroes of Heritability

189 **Having arrived in the Caribbean:** Brookes 2001.
190 **Between 1910 and 1937:** Brookes 2001.
190 **Others occupy more violent niches:** Marshall 2012.
191 **"Everything in modern genetics":** Brookes 2001, p. 7.
191 **"Many of our discoveries about inheritance":** Kelly Dyer, personal communication, October 15, 2018.
191 **Among the sweeping fruit fly–assisted:** Brookes 2001.
191 **The early 1980s saw the emergence:** Brookes 2001.
192 **Such is CRISPR's power:** Patrick O'Grady, personal communication, April 22, 2019.
192 **More surprising than that:** Brookes 2001; Yin et al. 1995.
192 **So far, fruit fly research:** Sverdrup-Thygeson 2019.
192 **There are about a hundred thousand:** Patrick O'Grady, personal communication, April 22, 2019.
192 **The diversity of mutations wrought:** Owald et al. 2015.
192 **The Ken and Barbie mutant:** Iyer 2015.
201 **"Mutants that disrupt many":** Sokolowski 2010, p. 790.
202 **Compare that with a 1945 book:** Fullaway and Krauss 1945.
203 **It's estimated that the first Hawaiian island:** Brookes 2001.
207 **It turns out that longer sperm:** Lüpold et al. 2016.
207 **To provide an illustration:** Patterson and Stone 1952.
209 **It goes to show that not all:** Lüpold et al. 2016.
210 **But two years later, when the last generation:** Brookes 2001.
210 **It is a measure of the adaptiveness:** Izutsu et al. 2015.

Chapter 10: Vectors and Pests

211 **"In the time it takes you to read":** Marshall 2012, p. 62.
211 **In his celebrated 1997 book:** Diamond 1997.
212 **Flies are credited with driving:** Winegard 2019.
212 **Thus, mosquitoes act as flying:** Spielman and D'Antonio 2002.
212 **Biting midges (Ceratopogonidae) are known:** Borkent 2005; Meiswinkel et al. 2004.
212 **Oroya fever, transmitted by sand flies:** Gaul 1953.
212 **Since the year 2000, mosquitoes:** Winegard 2019.
213 **That's about 52 billion out of 108 billion:** Winegard 2019.
213 **Until World War II, many more soldiers:** Grzimek 2003.
213 **Depending on whom you ask:** Cibulskis et al. 2016; Winegard 2019.
213 **In all, over 15 diseases are transmitted:** Winegard 2019.
213 ***Culex* mosquitoes are also culpable:** www.nationalgeographic.com/animals/invertebrates/group/mosquitoes/.
213 **Fortunately, there is no evidence:** Vandertogt 2020.
213 **The World Health Organization (WHO) reports:** World Health Organization 2018.
214 **malaria can progress:** Government of Canada 2016.
214 **When it comes into contact:** Consuelo et al. 2014.
214 **The parasite further manipulates:** Winegard 2019.
215 **By 1950, through a combination:** Spielman and D'Antonio 2002.
215 **The famous slave uprising:** Spielman and D'Antonio 2002.
215 **Between 1693 and 1905:** Patterson 1992.
216 **Pesticides are also hazardous to us:** Rifai 2017.
216 **Lastly, there is the ever-present specter:** Jeffries et al. 2018.
216 **The sterile male technique (SMT):** Rivers and Dahlem 2014.
217 **Considering that specific gene sequences:** Citations in Sun et al. 2017.
218 **This lowers the reproductive success:** Hancock et al. 2011.
218 **A trial in the northeastern Australian city:** Callaway 2018.
219 **While transgenic flies can reasonably:** Min et al. 2018.
219 **This strategy could allow pests:** Min et al. 2018.
219 **What if we could, for example:** Matthews et al. 2016, cited in Min et al. 2018.
219 **On the grimmer side:** Bier et al. 2018.
219 **Indeed, theoretical models have shown:** Sarkar 2018.

220 Over time, these sorts of mechanisms: Min et al. 2018.
220 A recent risk workshop: Roberts et al. 2017, cited in Min et al. 2018.
220 A cost-benefit analysis: Min et al. 2018.
220 As part of its Target Malaria effort: Sarkar 2018.
221 Evolution fights back with tactics: Sarkar 2018.
221 Experiments in multiple insect species: Sarkar 2018.
221 By the early 1990s: World Health Organization 1992.
221 And because the embattled parasite: Spielman and D'Antonio 2002.
222 In 2000, 10 percent of the world's: Spielman and D'Antonio 2002.
222 Resistance to its successor, mefloquine: Spielman and D'Antonio 2002.
222 Overall resistance to pesticides rose: MacNeal 2017.
222 The WHO's report for 2010 to 2016: World Health Organization 2020.
222 A 2016 clinical trial of Mosquirix: "Malaria Vaccine Loses Effectiveness over Several Years."
222 Overall efficacy across seven years: Olotu et al. 2016.
222 However, in a larger trial: RTS,S Clinical Trials Partnership 2015.
222 On the basis of evidence to date: RTS,S Clinical Trials Partnership 2015.
223 Sure enough, recent years have witnessed: Hoppé 2016.
223 Sometimes the most reliable methods: Spielman and D'Antonio 2002.
223 More than two thirds of this decline: Hoppé 2016.
223 Some mosquito species are adapting: Zivkovic 2012.
223 Dengue is the most important: Ferreira and Silva-Filha 2013.
223 Only nine countries: Dickie 2019.
224 In 2019, the Americas also saw: Cunningham 2019.
224 An Australian research team: Meyer et al. 2019.
225 Among its advantages, the card-based method: Milius 2019.
225 They were able to detect the DNA presence: Cook et al. 2017.
225 European clinics and hospitals: Winegard 2019.
225 In addition to dengue: McKie 2019.
226 Bees pollinating flowers: Grzimek 2003.
226 Only about 1 percent of insect species: Grzimek 2003.
226 Depending on what source you consult: Grzimek 2003; Hervé 2018.

226 Flies are significant contributors: Avis-Riordan 2019.
226 When we plant acres and acres of corn: Paulson and Eaton 2018.
227 It's worth bearing in mind: Muto et al. 2018.
227 The species is also spreading fast: Radonjić et al. 2019.
228 Tachinid flies represent a broad reservoir: Marshall 2012.
228 This approach has the time- and cost-saving advantage: Elkinton and Boettner 2005.
228 In a 2006 study: Losey and Vaughan 2006.
229 The fly is killing about 80 percent: Elkinton and Boettner 2005.
229 The three big problems of insecticides: Hervé 2018.
229 IPM's methods include: Lanouette et al. 2017.
229 In tests, tea tree oil, andiroba oil: Klauck et al. 2014.
229 an application of hemp oil: Benelli et al. 2018.
229 Perhaps because they are less profitable: Klauck et al. 2014.
230 A 2019 study found that both: Hodgdon et al. 2019.
230 Malaysia's carambola (star fruit) export industry: Ansari et al. 2012.
230 Excreted in manure, ivermectin: Berenbaum 2018, Floate 1998.
231 In the current era of global commerce: Van Niekerken 2018.
231 Nevertheless, with horticultural industries: Enkerlin et al. 2017.
232 As of 2017, screwworms: Whitworth.
232 Two hundred thousand stable flies emerge: Taylor et al. 2012.
233 A further drawback to pesticides: Paulson and Eaton 2018.
234 By those standards, shipping containers: Spielman and D'Antonio 2002.
234 We've known for half a century: Paine 1969.
235 Rachel Carson did more than anyone: Carson 1962.

Chapter 11: Detectives and Doctors

237 This dipteran detective device: Lauck 1998.
237 There is a whole suite of larvae: Grzimek 2003.
238 Case in point: the presence of active maggots: Martín-Vega et al. 2011.
239 It is enormously helpful: Rivers and Dahlem 2014.
239 Females of the flesh fly *Sarcophaga utilis*: Rivers and Dahlem 2014.

239 **PMI is a critical measure:** Rivers and Dahlem 2014.
239 **Entomologists use knowledge:** Benecke 1998.
240 **In all, hundreds of chemicals:** Rivers and Dahlem 2014.
240 **Other flies will burrow:** MacNeal 2017.
240 **A study sought to determine:** Bhadra et al. 2014.
241 **Though they can appear on remains:** Nazni et al. 2008.
241 **the presence of drugs:** Benecke 1998.
242 **The man had lain dead for some time:** Greenberg and Kunich 2002.
242 **I looked up Ruxton:** http://aboutforensics.co.uk/buck-ruxton/.
243 **"Think of a dead body":** Rivers and Dahlem 2014, p. 71.
245 **Most of the celebrated cases:** Sultan 2006.
246 **It also became the subject of a bestselling book:** LeBourdais 1966.
247 **A reexamination of the insect data:** VanLaerhoven et al. 2019.
249 **When she was identified:** Goff and Lord 1994.
249 **Subsequently, the detection of drugs:** http://courses.biology.utah.edu/feener/5445/Lecture/Bio5445%20Lecture%2026.pdf.
249 **The fact that flesh fly larvae develop faster:** Paulson and Eaton 2018.
249 **When researchers injected various concentrations:** Bourel et al. 2001.
250 **As yet, there are no university degree programs:** www.forensicscolleges.com/blog/resources/college-forensic-entomology-programs.
250 **Several new molecular techniques:** Rivers and Dahlem 2014.
251 **Presence or absence of a specific:** Thompson et al. 2013.
251 **On the other hand, experts caution:** Thompson et al. 2013.
252 **Experts are working to develop:** Rivers and Dahlem 2014.
253 **We've known about maggots' ability:** Gaydos 2016.
254 **A pharmaceutical company, Lederle Laboratories:** Sherman et al. 2013.
255 **As just one example, a 2012 study:** Arnold 2013.
256 **That converts to a pound of removal:** Sherman et al. 2013.
256 **Greenbottle fly maggots also release ammonia:** Deyrup 2005.
257 **Patient anxiety (the "yuck factor"):** Sherman et al. 2013.
259 **As of 2013, some 80,000 patients:** Sherman et al. 2013.

Chapter 12: Caring about Flies

263 but unlike a bomber, these flies carry: Marshall 2012.
264 Lockwood's argument is formalized: Brotton 2017.
264 "Experiments were conducted": De Silva et al. 2015.
265 Flies' sensitivity to stimuli normally: Khuong et al. 2019.
266 By consuming microorganisms: Waldbauer 2003.
266 Midges own gold medals: Hudson et al. 2012.
266 As winged adults, they are: Deyrup 2005; www.onthewingphotography.com/wings/2011/05/14/midges-and-birds-food-for-thought/.
266 A recent survey of midges in Canada: Hebert et al. 2016.
267 In his influential 2005 book: Louv 2005.
267 A few years earlier: Wandersee and Schussler 1999.
268 That's what happened to the Easter Islanders: Cartwright 2014.
268 In the preface to their 1983 book: Ehrlich and Ehrlich 1983.
269 According to the best available data: Sánchez-Bayo and Wyckhuys 2019.
269 *The New York Times*, in a somber: *The New York Times* editorial board 2017.
269 During this time, mean maximum temperatures: Lister and Garcia 2018.
269 David Wagner, an expert in invertebrate conservation: Guarino 2018.
270 Extrapolated to all of Illinois: McKenna et al. 2001, in Berenbaum 2018.
270 Densities of flies, beetles, bees, and wasps: McKenna et al. 2001.
270 So it goes with the insectivorous lizards: Rosenberg et al. 2019.
271 The ratios are essentially unchanged: Bar-On et al. 2018.
273 The group has a strict no-collection: McLean 2019.
274 "That's what you get for being nice to wildlife!": www.youtube.com/watch?v=VWHdYuUDh1Y.
274 Since 2014, actor Morgan Freeman: Nace 2019.
274 Freeman installed 26 hives: www.youtube.com/watch?v=N96aCa9mEgw.
275 If all this beneficence: Lauck 1998, p. 67.
275 There is evidence for an innate human fear: Hoehl et al. 2017.
275 Flowers, houseflies, and fishes: New and German 2015.
275 "We are born without the fear": Naskrecki 2005, p. 1.
276 "No animal or plant in nature": Franzen 2018, p. 251.

Bibliography

Abbott, Jessica. "Self-Medication in Insects: Current Evidence and Future Perspectives." *Ecological Entomology* 39, no. 3 (June 2014): 273–80. https://doi.org/10.1111/een.12110.

Adamo, Shelly Anne. "Do Insects Feel Pain? A Question at the Intersection of Animal Behaviour, Philosophy and Robotics." *Animal Behaviour* 118 (August 2016): 75–79. https://doi.org/10.1016/j.anbehav.2016.05.005.

Aihara, Ikkyu, Priyanka de Silva, and Ximena E. Bernal. "Acoustic Preference of Frog-Biting Midges (*Corethrella* spp) Attacking Túngara Frogs in their Natural Habitat." *Ethology* 122, no. 2 (2016): 105–13. doi:10.1111/eth.12452.

Alem Sylvain, et al. "Associative Mechanisms Allow for Social Learning and Cultural Transmission of String Pulling in an Insect." *PLOS Biology* 14, no. 10 (October 4, 2016). https://doi.org/10.1371/journal.pbio.1002564.

Alsan, Marcella. "The Effect of the Tsetse Fly on African Development." *American Economic Review* 105, no. 1 (January 2015): 382–410, https://doi.org/10.1257/aer.20130604.

Alupay, J. S., S. P. Hadjisolomou, and R. J. Crook. "Arm Injury Produces Long-Term Behavioral and Neural Hypersensitivity in Octopus." *Neuroscience Letters* 558 (2014): 137–42.

Angioy, A.-M., et al. "Function of the Heater: The Dead Horse Arum Revisited." *Proceedings of the Royal Society B: Biological Sciences* 271, supplement 3 (February 7, 2004): S13–S15. https://doi.org/10.1098/rsbl.2003.0111.

Ansari, Mohd Shafiq, Fazil Hasan, and Nadeem Ahmad. "Threats to Fruit and Vegetable Crops: Fruit Flies (Tephritidae): Ecology, Behaviour, and Management." *Journal of Crop Science and Biotechnology* 15 (2012): 169–88. https://doi.org/10.1007/s12892-011-0091-6.

Arbuthnott, Devin, et al. "Mate Choice in Fruit Flies Is Rational and Adaptive." *Nature Communications* 8 (2017): 13953. https://doi.org/10.1038/ncomms13953.

Armitage, P. D., P. S. Cranston, and L. C. V. Pinder. *The Chironomidae: Biology and Ecology of Non-Biting Midges*. London: Chapman & Hall, 1995.

Armstrong, Adrian J., and Andy Blackmore. "Tsetse Flies Should Remain in Protected Areas in KwaZulu-Natal." *Koedoe* 59, no. 1 (2017): a1432. https://doi.org/10.4102/koedoe.v59i1.1432.

Arnold, Carrie. "New Science Shows How Maggots Heal Wounds." *Scientific American*, April 1, 2013. www.scientificamerican.com/article/news-science-shows-how-maggots-heal-wounds/?redirect=1 (accessed June 3, 2019).

Avis-Riordan, Katie. "Ten Insect Pests That Threaten the World's Plants." Royal Botanical Gardens, Kew, March 20, 2019. www.kew.org/read-and-watch/insect-pests-biggest-threat-plants.

Bächtold, Alexandra, and Kleber Del-Claro. "Predatory Behavior of *Pseudodorus clavatus* (Diptera, Syrphidae) on Aphids Tended by Ants." *Revista Brasileira de Entomologia* 57, no. 4 (October–December 2013): 437–39. https://doi.org/10.1590/S0085-56262013005000030.

Baer, William S. "The Treatment of Chronic Osteomyelitis with the Maggot (Larva of the Blow Fly)." *Journal of Bone and Joint Surgery* (American volume), 13 (1931): 438–75.

Bänziger, Hans. "Pollination of a Flowering Oddity: *Rhizanthes zippelii* (Blume) Spach (Rafflesiaceae)." *Natural History Bulletin of the Siam Society* 44 (1996): 113–42.

Bar-On, Yinon M., Rob Phillips, and Ron Milo. "The Biomass Distribution on Earth." *Proceedings of the National Academy of Sciences of the United States of America* 115, no. 25 (June 19, 2018): 6506–11. https://doi.org/10.1073/pnas.1711842115.

Barron, Andrew B., and Colin Klein. "What Insects Can Tell Us about the Origins of Consciousness." *Proceedings of the National Academy of Sciences of the United States of America* 113, no. 18 (May 3, 2016): 4900–8. https://doi.org/10.1073/pnas.1520084113.

Barry, Dave. "Bug Off!" In *Insect Lives: Stories of Mystery and Romance from a Hidden World*, ed. Erich Hoyt and Ted Schultz, 46–48. New York: John Wiley & Sons, 1999.

Barth, Friedrich G. *Insects and Flowers: The Biology of a Partnership*. Princeton, NJ: Princeton University Press, 1985.

Barton-Browne, Lindsay B., Roger J. Bartell, and Harry H. Shorey. "Pheromone-Mediated Behaviour Leading to Group Oviposition in the Blowfly *Lucilia cuprina*." *Journal of Insect Physiology* 15 (1969): 1003–14.

Bateson, Melissa, et al. "Agitated Honeybees Exhibit Pessimistic Cognitive Biases." *Current Biology* 21, no. 12 (June 21, 2011): 1070–73. https://doi.org/10.1016/j.cub.2011.05.017.

Benecke, Mark. "Six Forensic Entomology Cases: Description and Commentary." *Journal of Forensic Science* 43, no. 4 (August 1998): 797–805.

Benelli, Giovanni, et al. "Contest Experience Enhances Aggressive Behaviour in a Fly: When Losers Learn to Win." *Scientific Reports* 5, article 9347 (March 20, 2015). https://doi.org/10.1038/srep09347.

———. "The Essential Oil from Industrial Hemp (*Cannabis sativa* L.) By-products as an Effective Tool for Insect Pest Management in Organic Crops." *Industrial Crops and Products* 122, no. 10 (October 15, 2018): 308–15. https://doi.org/10.1016/j.indcrop.2018.05.032.

Berenbaum, May. "Fly on the Wall." *American Entomologist* 49, no. 4 (Winter 2003): 196–97.

———. "Lords of the Flies: Insects, Humans, and the Fate of the World We Share." *The Common Reader: A Journal of the Essay* (January 4, 2018). https://commonreader.wustl.edu/c/lords-of-the-flies/ (accessed November 7, 2018).

Bernal, Ximena E., and Priyanka de Silva. "Cues Used in Host-Seeking Behavior by Frog-Biting Midges (*Corethrella* spp. Coquillet)." *Journal of Vector Ecology* 40, no. 1 (June 5, 2015). https://doi.org/10.1111/jvec.12140.

Bernal, Ximena E., A. Stanley Rand, and Michael J. Ryan. "Acoustic Preferences and Localization Performance of Blood-Sucking Flies (*Corethrella* Coquillett) to Túngara Frog Calls." *Behavioral Ecology* 17, no. 5 (September/October 2006): 709–15. https://doi.org/10.1093/beheco/arl003.

Bhadra, Parna, Andrew J. Hart, and Martin Jonathan Richard Hall. "Factors Affecting Accessibility to Blowflies of Bodies Disposed in Suitcases." *Forensic Science International* 239 (June 2014): 62–72. https://doi.org/10.1016/j.forsciint.2014.03.020.

Bier, Ethan, et al. "Advances in Engineering the Fly Genome with the CRISPR-Cas System." *Genetics* 208, no. 1 (January 1, 2018): 1–18. https://doi.org/10.1534/genetics.117.1113.

Blaj, Gabriel, and J. Hans van Hateren. "Saccadic Head and Thorax Movements in Freely Walking Blowflies." *Journal of Comparative Physiology A: Neuroethology, Sensory, Neural, and Behavioral Physiology* 190, no. 11 (November 2004): 861–68. https://doi.org/10.1007/s00359-004-0541-4.

Blake, William. "The Fly." In *Songs of Experience*, 1794. https://poets.org/poem/fly.

Bland, Alastair. "Is the Livestock Industry Destroying the Planet? For the Earth's Sake, Maybe It's Time We Take a Good, Hard Look at Our Dietary Habits." *Smithsonian*, August 1, 2012. www.smithsonianmag.com/travel/is-the-livestock-industry-destroying-the-planet-11308007/.

Bogarín, Diego, et al. "Pollination of *Trichosalpinx* (Orchidaceae: Pleurothallidinae) by Biting Midges (Diptera: Ceratopogonidae)." *Botanical Journal of the Linnean Society* 186, no. 4 (April 2018): 510–43. https://doi.org/10.1093/botlinnean/box087.

Boisvert, Michael J., and David F. Sherry. "Interval Timing by an Invertebrate, the Bumble Bee *Bombus impatiens*." *Current Biology* 16, no. 16 (August 22, 2006): 1636–40. https://doi.org/10.1016/j.cub.2006.06.064.

Bonham Carter, Violet. *Winston Churchill As I Knew Him*. London: Eyre & Spottiswoode, 1965. (Published in the United States as *Winston Churchill: An Intimate Portrait*.)

Borkent, Art. "The Biting Midges, the Ceratopogonidae (Diptera)." In *Biology of Disease Vectors*, ed. W. C. Marquardt. San Diego: Elsevier Academic Press, 2005.

———. *Biting Midges in the Cretaceous Amber of North America (Diptera: Ceratopogonidae)*. Leiden: Backhuys, 1995.

———. "The Frog-Biting Midges of the World (Corethrellidae: Diptera)." *Zootaxa* 1804, no. 1 (June 16, 2008): 1–456. https://doi.org/10.11646/zootaxa.1804.1.1.

Borkent, Art, and John Bissett. "Gall Midges (Diptera: Cecidomyiidae) Are Vectors for Their Fungal Symbionts." *Symbiosis* 1 (1985): 185–94.

Borkent, Art, and Patrycja Dominiak. *Catalog of the Biting Midges of the World (Diptera: Ceratopogonidae)*. Auckland: Magnolia Press, 2020.

Borkent, Art, et al. "Remarkable Fly (Diptera) Diversity in a Patch of Costa Rican Cloud Forest: Why Inventory Is a Vital Science." *Zootaxa* 4402, no. 1 (March 27, 2018): 53–90. https://doi.org/10.11646/zootaxa.4402.1.3.

Bosch, Guido, et al. "Standardisation of Quantitative Resource Conversion Studies with Black Soldier Fly Larvae." *Journal of Insects as Food and Feed* 6, no. 2 (August 27, 2019, online): 95–109. https://doi.org/10.3920/JIFF2019.0004.

Bourel, Benoit, et al. "Morphine Extraction in Necrophagous Insects Remains for Determining Ante-Mortem Opiate Intoxication." *Forensic Science International* 120, no. 1–2 (August 15, 2001): 127–31. https://doi.org/10.1016/s0379-0738(01)00428-5.

Bragança, Marcos Antonio Lima, et al. "Phorid Flies Parasitizing Leaf-Cutting Ants: Their Occurrence, Parasitism Rates, Biology and the First Account of Multiparasitism." *Sociobiology* 63, no. 4 (2016): 1015–21. https://doi.org/10.13102/sociobiology.v63i4.1077.

Briceño, R. Daniel, and William Eberhard. "Copulatory Dialogues between Male and Female Tsetse Flies (Diptera: Muscidae: *Glossina pallidipes*)." *Journal of Insect Behavior* 30 (2017): 394–408. https://doi.org/10.1007/s10905-017-9625-1.

———. "Species-Specific Behavioral Differences in Tsetse Fly Genital Morphology and Probable Cryptic Female Choice." In *Cryptic Female Choice in Arthropods*, ed. A. V. Peretti and A. Aisenberg. Cham, Switzerland: Springer International, 2015.

———, and Alan S. Robinson. "Copulation Behaviour of *Glossina pallidipes* (Diptera: Muscidae) outside and inside the Female, with a Discussion of Genitalic Evolution." *Bulletin of Entomological Research* 97, no. 5 (October 2007): 471–88. https://doi.org/10.1017/S0007485307005214.

Brockmann, H. Jane. "Tool Using in Wasps." *Psyche* 92 (1985): 309–29.

Brookes, Martin. *Fly: The Unsung Hero of 20th-Century Science*. New York: Ecco, 2001.

Brotton, Melissa J., ed. "Ecotheology and Nonhuman Ethics in Society: A Community of Compassion." Lanham, MD: Lexington Books, 2017.

Brown, Brian V., and Donald H. Feener, Jr. "Efficiency of Two Mass Sampling Methods for Sampling Phorid Flies (Diptera: Phoridae) in a Tropical Biodiversity Survey." *Contributions in Science* 459 (1995): 1–10.

Brown, Brian V., Giar-Ann Kung, and Wendy Porras. "A New Type of Ant-Decapitation in the Phoridae (Insecta: Diptera)." *Biodiversity Data Journal* 3 (2015). https://bdj.pensoft.net/article/4299 (accessed December 1, 2017).

Brown, Elizabeth Nolan. "Sexual Frustration Is Bad for Your Health." *Bustle*, December 3, 2013. https://www.bustle.com/articles/9879-sexual-frustration

-can-be-bad-for-your-health-and-thats-not-just-a-pickup-line (accessed August 13, 2020).

Brunel, Odette, and Juan Rull. "The Natural History and Unusual Mating Behavior of *Euxesta bilimeki* (Diptera: Ulidiidae)." *Annals of the Entomological Society of America* 103, no. 1 (January 1, 2010): 111–19. https://doi.org/10.1093/aesa/103.1.111.

Byron, Ellen. "Bugs, the New Frontier in House-Cleaning." *The Wall Street Journal*, July 15, 2017.

Callaway, Ewen. "Dengue Rates Plummet in Australian City After Release of Modified Mosquitoes: Insects Were Deliberately Infected with Bacteria That Interrupt Transmission of the Disease." *Nature*, August 8, 2018. https://doi.org/10.1038/d41586-018-05914-3.

Cammaerts, Marie-Claire, and Roger Cammaerts. "Are Ants (Hymenoptera, Formicidae) Capable of Self Recognition?" *Journal of Science* 5, no. 7 (2015): 521–32.

Camp, Jeremy Vann. "Host Attraction and Host Selection in the Family Corethrellidae (Wood and Borkent) (Diptera)." MS thesis, Georgia Southern University, Statesboro, 2006.

Card, Gwyneth, and Michael H. Dickinson. "Visually Mediated Motor Planning in the Escape Response of *Drosophila*." *Current Biology* 18, no. 17 (September 9, 2008): 1300–7. https://doi.org/10.1016/j.cub.2008.07.094.

Carson, Rachel. *Silent Spring*. Boston: Houghton Mifflin, 1962.

Cartwright, Mark. "The Classic Maya Collapse." *Ancient History Encyclopedia*, October 18, 2014. www.ancient.eu/article/759/the-classic-maya-collapse/.

Cell Press. "To Kill Off Parasites, an Insect Self-Medicates with Alcohol." *ScienceDaily*, February 16, 2012. www.sciencedaily.com/releases/2012/02/120216133428.htm.

Chinery, Michael. *Amazing Insects: Images of Fascinating Creatures*. Richmond Hill, Ontario: Firefly Books, 2008.

Cibulskis, Richard E., et al. "Malaria: Global Progress 2000–2015 and Future Challenges." *Infectious Diseases of Poverty* 5, no. 61 (June 2016). https://doi.org/10.1186/s40249-016-0151-8.

Clastrier, Jean, Daniel Grand, and Jean Legrand. "Observations exceptionnelles en France de *Forcipomyia* (*Pterobosca*) *paludis* (Macfie), parasite des ailes de Libellules (Diptera, Ceratopogonidae et Odonata)." *Bulletin de la Société Entomologique de France* 99, no. 2 (June 1994): 127–30. www.persee.fr/doc/bsef_0037-928x_1994_num_99_2_17051.

Coatsworth, John, et al. *Global Connections: Politics, Exchange, and Social Life in World History*. Vol. 1, *To 1500*. Cambridge, UK: Cambridge University Press, 2015.

Coen, Philip, et al. "Sensorimotor Transformations Underlying Variability in Song Intensity during *Drosophila* Courtship." *Neuron* 89, no. 3 (February 3, 2016): 629–44. https://doi.org/10.1016/j.neuron.2015.12.035.

Coghlan, Andy. "Male Fruit Flies Feel Pleasure When They Ejaculate." *New Scientist*, April 19, 2018. www.newscientist.com/article/2166889-male-fruit-flies-feel-pleasure-when-they-ejaculate/.

Colpaert, F. C., et al. "Self-Administration of the Analgesic Suprofen in Arthritic Rats: Evidence of *Mycobacterium butyricum*–Induced Arthritis as an Experimental Model of Chronic Pain." *Life Sciences* 27 (1980): 921–28.

Committee Opinion No. 589. "Female Age-Related Fertility Decline." *Fertility and Sterility* 101 (March 2014): 633–34. https://doi.org/10.1016/j.fertnstert.2013.12.03.

Cook, Darren A. N., et al. "A Superhydrophobic Cone to Facilitate the Xenomonitoring of Filarial Parasites, Malaria, and Trypanosomes Using Mosquito Excreta/Feces." *Gates Open Research* 1, no. 7 (November 6, 2017). https://doi.org/10.12688/gatesopenres.12749.1; (April 27, 2018). https://doi.org/10.12688/gatesopenres.12749.2.

Coolen, Isabelle, Olivier Dangles, and Jérôme Casas. "Social Learning in Noncolonial Insects?" *Current Biology* 15, no. 21 (November 8, 2005): 1931–35. https://doi.org/10.1016/j.cub.2005.09.015.

Cousins, Melanie, et al. "Modelling the Transmission Dynamics of *Campylobacter* in Ontario, Canada, Assuming House Flies, *Musca domestica*, Are a Mechanical Vector of Disease Transmission." *Royal Society Open Science* 6, no. 2 (February 13, 2019). https://doi.org/10.1098/rsos.181394.

Cross, Fiona R., and Robert R. Jackson. "The Execution of Planned Detours by Spider-Eating Predators." *Journal of the Experimental Analysis of Behavior* 105, no. 1 (January 2016): 194–210. https://doi.org/10.1002/jeab.189.

Cunningham, Aimee. "Dengue Cases in the Americas Have Reached an All-Time High." *Science News*, November 20, 2019. www.sciencenews.org/article/dengue-cases-americas-have-reached-all-time-high.

Curic, Goran, et al. "Identification of Person and Quantification of Human DNA Recovered from Mosquitoes (Culicidae)." *Forensic Science International: Genetics* 8, no. 1 (January 1, 2014): 109–12. https://doi.org/10.1016/j.fsigen.2013.07.011.

Curran, Charles Howard. *The Families and Genera of North American Diptera*, 2nd ed. Woodhaven, NY: Henry Tripp, 1965.

Dacke, Marie, et al. "Dung Beetles Use the Milky Way for Orientation." *Current Biology* 23, no. 4 (February 18, 2013): 298–300, https://doi.org/10.1016/j.cub.2012.12.034.

Danbury, T. C., et al. "Self-Selection of the Analgesic Drug, Carprofen, by Lame Broiler Chickens." *Veterinary Record* 146 (2000): 307–11.

Dason, Jeffrey S., et al. "*Drosophila melanogaster* Foraging Regulates a Nociceptive-like Escape Behavior through a Developmentally Plastic Sensory Circuit." *Proceedings of the National Academy of Sciences*. June 18, 2019. https://doi.org/10.1073/pnas.1820840116.

Davidoff, Ken, and George A. King III. "The Night When Bugs Changed the Course of Yankees History." *New York Post*, October 4, 2017. https://nypost.com/2017/10/04/the-night-when-bugs-changed-the-course-of-yankees-history/.

Dawkins, Marian S. *Animal Suffering: The Science of Animal Welfare*. New York: Chapman & Hall, 1980.

De Moraes, Consuelo M., et al. "Malaria-Induced Changes in Host Odors Enhance Mosquito Attraction." *Proceedings of the National Academy of Sciences of the United States of America* 111, no. 30 (July 29, 2014): 11079–84. https://doi.org/10.1073/pnas.1405617111.

De Morgan, Augustus. *A Budget of Paradoxes*. London: Longmans, Green, 1872. www.maa.org/press/periodicals/convergence/mathematical-treasure-de-morgan-s-budget-of-paradoxes.

de Silva, Priyanka, Brian Nutter, and Ximena E. Bernal. "Use of Acoustic Signals in Mating in an Eavesdropping Frog-Biting Midge." *Animal Behaviour* 103 (May 2015): 45–51. https://doi.org/10.1016/j.anbehav.2015.02.002.

Deora, Tanvi, Amit Kumar Singh, and Sanjay P. Sane. "Biomechanical Basis of Wing and Haltere Coordination in Flies." *Proceedings of the National Academy of Sciences of the United States of America* 112, no. 5 (January 2015): 1481–86. https://doi.org/10.1073/pnas.1412279112.

Deyrup, Mark, and Thomas C. Emmel. *Florida's Fabulous Insects*. Hawaiian Gardens, CA: World Publications, 1999.

Diamond, Jared M. *Guns, Germs, and Steel: The Fates of Human Societies*. New York: W. W. Norton, 1997.

Dickie, Gloria. "Nepal Is Reeling from an Unprecedented Dengue Outbreak." *Science News*, October 7, 2019. www.sciencenews.org/article/nepal-reeling-from-unprecedented-dengue-virus-outbreak.

Disney, R. H. L. *Scuttle Flies: The Phoridae*. London: Chapman & Hall, 1994.

Dodson, Calaway H. "The Importance of Pollination in the Evolution of the Orchids of Tropical America." *American Orchid Society Bulletin* 31, no. 9 (September 1962): 641–735.

Doherty, Mark. "Bug-Growing Facility Will Buzz into Balzac Next Spring." *StarMetro* (Calgary), September 4, 2018.

Dowling, Stephen. "Do Mosquitoes Feel the Effects of Alcohol?" BBC Future, March 13, 2019. www.bbc.com/future/story/20190313-will-mosquitoes-bite-me-more-when-ive-been-drinking.

Downes, J. A. "Feeding and Mating in the Insectivorous Ceratopogoninae (Diptera)." *Memoirs of the Entomological Society of Canada*, 104 (1978).

du Plessis, Marc, et al. "Pollination of the 'Carrion Flowers' of an African Stapeliad (*Ceropegia mixta*: Apocynaceae): The Importance of Visual and Scent Traits for the Attraction of Flies." *Plant Systematics and Evolution* 304, no. 3 (March 2018): 357–72. https://doi.org/10.1007/s00606-017-1481-0.

Dyer, Adrian G., Christa Neumeyer, and Lars Chittka. "Honeybee (*Apis mellifera*) Vision Can Discriminate between and Recognise Images of Human Faces." *Journal of Experimental Biology* 208, part 24 (December 2005): 4709–14. https://doi.org/10.1242/jeb.01929.

Ehrlich, Paul, and Anne Ehrlich. *Extinction: The Causes and Consequences of the Disappearance of Species*. New York: Ballantine Books, 1983.

Eisemann, C. H., et al. "Do Insects Feel Pain? A Biological View." *Experientia* 40, no. 2 (1984): 164–67.

Elkinton, Joe S., and George H. Boettner. "The Effects of *Compsilura concinnata*, an Introduced Generalist Tachinid, on Non-Target Species in North

America: A Cautionary Tale." In *Assessing Host Ranges for Parasitoids and Predators Used for Classical Biological Control: A Guide to Best Practice*, ed. Roy G. Van Driesche, Tara J. Murray, and Richard Reardon, 4–14. Washington, DC: US Department of Agriculture, 2005.

Enkerlin Hoeflich, Walther Raúl, et al. "The Moscamed Regional Programme: Review of a Success Story of Area-Wide Sterile Insect Technique Application." *Entomologia Experimentalis et Applicata*, Special Issue—Sterile Insect Technique, 164, no. 3 (September 19, 2017): 188–203.

Evans, Harold Ensign. "The Lovebug." In *The Pleasures of Entomology. Portraits of Insects and the People Who Study Them*. Washington, DC: Smithsonian Institution Press, 1985.

Farndon, John, Barbara Taylor, and Jen Green. *Bugs & Minibeasts: Beetles, Bugs, Butterflies, Moths, Insects, Spiders*. Illustrated Wildlife Encyclopedia series. London: Armadillo, 2014.

Farnham, Alex. "How We Benefit from Flies." *Animalist News*, March 28, 2014. www.youtube.com/watch?v=LxjbbNMyTMA&feature=youtu.be (accessed September 2, 2018).

Feener, Donald H., Jr., and Karen A. G. Moss. "Defense against Parasites by Hitchhikers in Leaf-Cutting Ants: A Quantitative Assessment." *Behavioural Ecology and Sociobiology* 26, no. 1 (January 1990): 17–29. https://doi.org/10.1007/BF00174021.

Ferreira, Lígia Maria, and Maria Helena Neves Lobo Silva-Filha. "Bacterial Larvicides for Vector Control: Mode of Action of Toxins and Implications for Resistance." *Biocontrol Science and Technology* 23, no. 10 (2013): 1137–68. https://doi.org/10.1080/09583157.2013.822472.

Floate, Kevin D. "Off-Target Effects of Ivermectin on Insects and on Dung Degradation in Southern Alberta, Canada." *Bulletin of Entomological Research* 88, no. 1 (February 1998): 25–35. https://doi.org/10.1017/S0007485300041523.

Food and Agriculture Organization of the United Nations. *Livestock's Long Shadow: Environmental Issues and Options*. Rome: FAO, 2006. www.fao.org/3/a-a0701e.pdf.

———. *The State of World Fisheries and Aquaculture: Opportunities and Challenges*. Rome: FAO, 2014. www.fao.org/3/a-i3720e.pdf.

Förster, Maria, Rolf G. Beutel, and Katharina Schneeberg. "Catching Prey with the Antennae: The Larval Head of *Corethrella appendiculata* (Diptera: Corethrellidae)." *Arthropod Structure & Development* 45, no. 6 (November 2016): 594–610. https://doi.org/10.1016/j.asd.2016.09.003.

Franzen, Jonathan. *The End of the End of the Earth: Essays*. New York: Farrar, Straus and Giroux, 2018.

Frauca, Harry. *Australian Insect Wonders*. Adelaide: Rigby, 1968.

Fullaway, David Timmins, and Noel Louis Hilmer Krauss. *Common Insects of Hawaii*. Honolulu: Tongg, 1945.

Gaimari, Stephen D., and Jim O'Hara. "C. P. Alexander Award." *Fly Times* 58 (April 2017): 1–2.

Gallup, Gordon G., Jr. "Chimpanzees: Self Recognition." *Science* 167 (1970): 86–87.

Galluzzo, Gary. "A Fly's Hearing: UI Study Shows Fruit Fly Is Ideal Model to Study Hearing Loss in People." *Iowa Now*, September 2, 2013. https://now.uiowa.edu/2013/09/flys-hearing.

Gardner, Elliot M., et al. "A Flower in Fruit's Clothing: Pollination of Jackfruit (*Artocarpus heterophyllus*, Moraceae) by a New Species of Gall Midge, *Clinodiplosis ultracrepidata* sp. nov. (Diptera: Cecidomyiidae)." *International Journal of Plant Sciences* 179, no. 5 (June 2018): 350–67. https://doi.org/10.1086/697115.

Gaul, Albro Tilton. *The Wonderful World of Insects*. New York: Rinehart, 1953.

Gaydos, Jaclyn. "History of Wound Care: Maggots: An Extraordinary Natural Phenomenon." *Today's Wound Clinic* 10, no. 4 (April 2016). www.todayswoundclinic.com/articles/history-wound-care-maggots-extraordinary-natural-phenomenon.

Gibson, William T., et al. "Behavioral Responses to a Repetitive Visual Threat Stimulus Express a Persistent State of Defensive Arousal in *Drosophila*." *Current Biology* 25, no. 11 (June 1, 2015): 1401–15. https://doi.org/10.1016/j.cub.2015.03.058.

Gilbert, Irwin H., Harry K. Gouck, and Nelson Smith. "Attractiveness of Men and Women to *Aedes aegypti* and Relative Protection Time Obtained with Deet." *Florida Entomologist* 49, no. 1 (March 1966): 53–66. https://doi.org/10.2307/3493317.

Giurfa, Martin. "Cognition with Few Neurons: Higher-Order Learning in Insects." *Trends in Neurosciences* 36, no. 5 (May 1, 2013): 285–94. https://doi.org/10.1016/j.tins.2012.12.011.

Goff, M. Lee, and Wayne D. Lord. "Entomotoxicology: A New Area for Forensic Investigation." *The American Journal of Forensic Medicine and Pathology* 15, no. 1 (March 1994): 51–57.

Goodall, Jane. "Learning from the Chimpanzees: A Message Humans Can Understand." *Science* 282, no. 5397 (December 18, 1998): 2184–85. https://doi.org/10.1126/science.282.5397.2184.

Göpfert, Martin C., and Daniel Robert. "Nanometre-Range Acoustic Sensitivity in Male and Female Mosquitoes." *Proceedings of the Royal Society B: Biological Sciences* 267, no 1442 (March 7, 2000): 453–57. https://doi.org/10.1098/rspb.2000.1021.

Gorman, James. "Trillions of Flies Can't All Be Bad." *The New York Times*, November 13, 2017. www.nytimes.com/2017/11/13/science/flies-biology.html.

Goulson, Dave, et al. "Predicting Calyptrate Fly Populations from the Weather, and Probable Consequences of Climate Change." *Journal of Applied Ecology* 42, no. 5 (September 2005): 795–804. https://doi.org/10.1111/j.1365-2664.2005.01078.x.

Government of Canada. "Symptoms of Malaria." Last updated April 21, 2016. www.canada.ca/en/public-health/services/diseases/malaria/symptoms-malaria.html.

Grafe, T. Ulmar, et al. "Studying the Sensory Ecology of Frog-Biting Midges (Corethrellidae: Diptera) and Their Frog Hosts Using Ecological Interaction Networks." *Journal of Zoology* 307, no. 1 (January 2019): 17–27. https://doi.org/10.1111/jzo.12612.

Grassberger, Martin, et al., eds. *Biotherapy—History, Principles and Practice: A Practical Guide to the Diagnosis and Treatment of Disease Using Living Organisms*. Dordrecht, Netherlands: Springer, 2013. https://doi.org/10.1007/978-94-007-6585-6.

Greenberg, Bernard, and John Charles Kunich. *Entomology and the Law: Flies as Forensic Indicators*. Cambridge, UK: Cambridge University Press, 2002.

Griffin, Donald R. *Animal Minds: Beyond Cognition to Consciousness*. Chicago: University of Chicago Press, 1992.

———. *The Question of Animal Awareness: Evolutionary Continuity of Mental Experience*. New York: Rockefeller University Press, 1981.

Griggs, Mary Beth. "Mosquitoes Learn Not to Mess with You When You Swat Them: And They'll Likely Go Looking for a Less Combative Meal." *Popular Science*, January 25, 2018. www.popsci.com/mosquitoes-probably-remember-when-you-try-to-swat-them (accessed January 29, 2019).

Groening, Julia, Dustin Venini, and Mandyam V. Srinivasan. "In Search of Evidence for the Experience of Pain in Honeybees: A Self-Administration Study." *Scientific Reports* 7, article 45825 (April 4, 2017). https://doi.org/10.1038/srep45825.

Grover, Dhruv, Takeo Katsuki, and Ralph J. Greenspan. "Flyception: Imaging Brain Activity in Freely Walking Fruit Flies." *Nature Methods* 13 (2016): 569–72.

Grover, Dhruv, et al. "Imaging Brain Activity during Complex Social Behaviors in *Drosophila* with Flyception2." *Nature Communications* 11, no. 623 (2020).

Grzimek, Don Bernhard, *Grzimek's Animal Life Encyclopedia*, 2nd ed. Vol. 3, *Insects*, ed. Michael Hutchins et al. Farmington Hills, MI: Gale Group, 2003.

Guarino, Ben. "'Hyperalarming' Study Shows Massive Insect Loss." *The Washington Post*, October 15, 2018. www.washingtonpost.com/science/2018/10/15/hyperalarming-study-shows-massive-insect-loss/?noredirect=on&utm_term=.75a1f83e2ab3.

———. "Watch These Bizarre Flies Dive Underwater Using Bubbles Like Scuba Suits." *The Washington Post*, November 20, 2017. www.washingtonpost.com/news/speaking-of-science/wp/2017/11/20/these-bizarre-flies-wear-bubbles-like-scuba-suits-to-dive-in-a-toxic-lake/?utm_term=.372cf575a632.

Halfwerk, Wouter, et al. "Adaptive Changes in Sexual Signalling in Response to Urbanization." *Nature Ecology & Evolution* 3, no. 3 (March 2019): 374–80. https://doi.org/10.1038/s41559-018-0751-8.

Hall, Andrew Brantley, et al. "A Male-Determining Factor in the Mosquito *Aedes aegypti*." *Science* 348, no. 6240 (June 12, 2015): 1268–70. https://doi.org/10.1126/science.aaa2850.

Hancock, Penelope A., Steven P. Sinkins, and H. Charles J. Godfray. "Strategies for Introducing *Wolbachia* to Reduce Transmission of Mosquito-Borne

Diseases." *PLOS Neglected Tropical Disease* 5, no. 4 (April 26, 2011). https://doi.org/10.1371/journal.pntd.0001024.

Hebert, Paul D. N., et al. "Counting Animal Species with DNA Barcodes: Canadian Insects." *Philosophical Transactions of the Royal Society B: Biological Sciences* 371, no. 1702 (September 5, 2016): 10. https://doi.org/10.1098/rstb.2015.0333.

Heid, Matt. "How Many Mosquito Bites Would It Take to Kill You (and Other Mosquito Musings)." Be Outdoors: Appalachian Mountain Club, July 1, 2014. www.outdoors.org/articles/amc-outdoors/how-many-mosquito-bites-would-kill-you.

Heinrich, Bernd. *Life Everlasting: The Animal Way of Death*. Boston: Houghton Mifflin Harcourt, 2012.

———. *Winter World: The Ingenuity of Animal Survival*. New York: Ecco, 2003.

Heisenberg, Martin, Reinhard Wolf, and Björn Brembs. "Flexibility in a Single Behavioral Variable of *Drosophila*." *Learning & Memory* 8, no. 1 (January–February 2001): 1–10.

Hervé, Maxime R. "Breeding for Insect Resistance in Oilseed Rape: Challenges, Current Knowledge and Perspectives." *Plant Breeding* 137, no. 1 (February 2018): 27–34. https://doi.org/10.1111/pbr.12552.

Hodgdon, Elisabeth A., et al. "Racemic Pheromone Blends Disrupt Mate Location in the Invasive Swede Midge, *Contarinia nasturtii*." *Journal of Chemical Ecology* 45, no. 7 (July 2019): 549–58. https://doi.org/10.1007/s10886-019-01078-0.

Hoehl, Stefanie, et al. "Itsy Bitsy Spider . . . : Infants React with Increased Arousal to Spiders and Snakes." *Frontiers in Psychology*, October 18, 2017. https://doi.org/10.3389/fpsyg.2017.01710 (accessed May 14, 2020).

Hoffmann, Constanze, et al. "Assessing the Feasibility of Fly-Based Surveillance of Wildlife Infectious Diseases." *Scientific Reports* 6, article 37952 (November 30, 2016). https://doi.org/10.1038/srep37952.

Hoppé, Mark. "Insecticide Resistance: Are We Losing the Battle to Control the Mosquito Vectors of Malaria?" *Outlooks on Pest Management* 27, no. 3 (June 2016): 116–19. https://doi.org/10.1564/v27_jun_05.

Howard, Leland O. *The Insect Book*. New York: Doubleday, Page, 1905.

Howard, Scarlett R., et al. "Numerical Ordering of Zero in Honey Bees." *Science* 360, no. 6393 (June 8, 2018): 1124–26. https://doi.org/10.1126/science.aar4975.

Hoyle, Graham. "Cellular Mechanisms Underlying Behavior—Neuroethology." *Advances in Insect Physiology* 7 (1970) 349–444. https://doi.org/10.1016/S0065-2806(08)60244-1.

Hudson, John, Katherine Hocker, and Robert H. Armstrong. *Aquatic Insects in Alaska*. Juneau: Nature Alaska Images, 2012.

Hussein, Mahmoud, et al. "Sustainable Production of Housefly (*Musca domestica*) Larvae as a Protein-Rich Feed Ingredient by Utilizing Cattle Manure." *PLOS One* 12, no. 2 (February 7, 2017). https://doi.org/10.1371/journal.pone.0171708.

Inouye, David W., et al. "Flies and Flowers III: Ecology of Foraging and Pollination." *Journal of Pollination Ecology* 16, no. 16 (2015): 115–33. www.pollinationecology.org/index.php?journal=jpe&page=article&op=view&path%5B%5D=333.

Iyer, Shruti. "14 of the Funniest Fruit Fly Gene Names." Bitesize Bio, March 2, 2015. https://bitesizebio.com/23221/14-of-the-funniest-fruit-fly-gene-names/.

Izutsu, Minako, et al. "Dynamics of Dark-Fly Genome under Environmental Selections." *Genes, Genomes, Genetics* 6, no. 2 (December 4, 2015): 365–76. https://doi.org/10.1534/g3.115.023549.

Jeffries, Claire L., Matthew E. Rogers, and Thomas Walker. "Establishment of a Method for *Lutzomyia longipalpis* Sand Fly Egg Microinjection: The First Step towards Potential Novel Control Strategies for Leishmaniasis," version 2. *Wellcome Open Research* 3 (August 2018): 55, https://doi.org/10.12688/wellcomeopenres.14555.2.

Jürgens, Andreas, and Adam Shuttleworth. "Carrion and Dung Mimicry in Plants." In *Carrion Ecology, Evolution, and Their Applications*, ed. M. Eric Benbow, Jeffery K. Tomberlin, and Aaron M. Tarone, 361–87. Boca Raton, FL: CRC Press, 2015.

Jürgens, Andreas, et al. "Chemical Mimicry of Insect Oviposition Sites: A Global Analysis of Convergence in Angiosperms." *Ecology Letters* 16 (2013): 1157–67.

Kacsoh, Balint Z., et al. "Fruit Flies Medicate Offspring after Seeing Parasites." *Science* 339, no. 6122 (February 22, 2013): 947–50. https://doi.org/10.1126/science.1229625.

Kain, Jamey S., Chris Stokes, and Benjamin L. de Bivort. "Phototactic Personality in Fruit Flies and Its Suppression by Serotonin and White." *Proceedings of the National Academy of Sciences of the United States of America* 109, no. 48 (November 27, 2012): 19834–39. https://doi.org/10.1073/pnas.1211988109.

Keller, Andreas. "A Cultural and Natural History of the Fly." *PLOS Biology* 5, no. 5 (May 15, 2007). https://doi.org/10.1371/journal.pbio.0050135.

Kern, Roland, Johannes Hans van Hateren, and Martin Egelhaaf. "Representation of Behaviourally Relevant Information by Blowfly Motion-Sensitive Visual Interneurons Requires Precise Compensatory Head Movements." *Journal of Experimental Biology* 209, no. 7 (April 1, 2006): 1251–60. https://doi.org/10.1242/jeb.02127.

Khuong, Thang M., et al. "Nerve Injury Drives a Heightened State of Vigilance and Neuropathic Sensitization in *Drosophila*." *Science Advances* 5, no. 7 (July 10, 2019). https://doi.org/10.1126/sciadv.aaw4099.

Kiderra, Inga. "First Peek into the Brain of a Freely Walking Fruit Fly." UC San Diego News Center, May 16, 2016. https://ucsdnews.ucsd.edu/pressrelease/first_peek_into_the_brain_of_a_freely_walking_fruit_fly.

Klauck, V., et al. "Insecticidal and Repellent Effects of Tea Tree and Andiroba Oils on Flies Associated with Livestock." *Medical and Veterinary Entomology* 28, supplement 1 (August 2014): 33–39. https://doi.org/10.1111/mve.12078.

Klein, Barrett A. "Insects in Art." In *Encyclopedia of Human-Animal Relationships: A Global Exploration of Our Connections with Animals*, ed. Marc Bekoff. Westport, CT: Greenwood Press, 2007, 92–99.

Klein, Colin, and Andrew B. Barron. "Insects Have the Capacity for Subjective Experience." *Animal Sentience* 9, no. 1 (2016). https://animalstudies repository.org/animsent/vol1/iss9/1/.

Knop, Eva, et al. "Rush Hours in Flower Visitors over a Day–Night Cycle." *Insect Conservation and Diversity* 11, no. 3 (May 2018): 267–75. https://doi .org/10.1111/icad.12277.

Krashes, Michael J., et al. "A Neural Circuit Mechanism Integrating Motivational State with Memory Expression in *Drosophila*." *Cell* 139, no. 2 (October 16, 2009): 416–27. https://doi.org/10.1016/j.cell.2009.08.035.

Lanouette, Geneviève, et al. "The Sterile Insect Technique for the Management of the Spotted Wing *Drosophila, Drosophila suzukii*: Establishing the Optimum Irradiation Dose." *PLOS One* 12, no. 9 (September 28, 2017). https://doi.org/10.1371/journal.pone.0180821.

Lauck, Joanne Elizabeth. *The Voice of the Infinite in the Small: Revisioning the Insect-Human Connection*. Mill Spring, NC: Swan, Raven, 1998.

LeBourdais, Isabel. *The Trial of Steven Truscott*. Toronto: McClelland & Stewart, 1966.

Lefebvre, Vincent, et al. "Are Empidine Dance Flies Major Flower Visitors in Alpine Environments? A Case Study in the Alps, France." *Biology Letters* 10, no. 11 (November 1, 2014). https://doi.org/10.1098/rsbl.2014.0742.

Le Neindre, Pierre, et al. "Animal Consciousness." *European Food Safety Authority Supporting Publications* 14, no. 4 (April 2017). https://doi.org/10.2903 /sp.efsa.2017.EN-1196.

Lim, T. M. "Production, Germination and Dispersal of Basidiospores of *Ganoderma pseudoferreum* on Hevea." *Journal of the Rubber Research Institute of Malaysia* 25, no. 2 (1977): 93–99.

Linger, Rebecca J., et al. "Towards Next Generation Maggot Debridement Therapy: Transgenic *Lucilia sericata* Larvae That Produce and Secrete a Human Growth Factor." *BMC Biotechnology* 16, article 30 (2016). https:// doi.org/10.1186/s12896-016-0263-z.

Lister, Bradford C., and Andres Garcia. "Climate-Driven Declines in Arthropod Abundance Restructure a Rainforest Food Web." *Proceedings of the National Academy of Sciences of the United States of America* 115, no. 44 (October 15, 2018): E10397–E10406. https://doi.org/10.1073/pnas.1722477115.

Lockwood, Jeff. *The Infested Mind: Why Humans Fear, Loathe, and Love Insects*. Oxford: Oxford University Press, 2014.

Losey, John E., and Mace Vaughan. "The Economic Value of Ecological Services Provided by Insects." *BioScience* 56, no. 4 (April 1, 2006): 311–23. https://doi.org/10.1641/0006-3568(2006)56[311:TEVOES]2.0.CO;2.

Louv, Richard. *Last Child in the Woods: Saving Our Children from Nature-Deficit Disorder*. Chapel Hill, NC: Algonquin Books, 2005.

Low, Tim. *The New Nature: Winners and Losers in Wild Australia*. Victoria, Australia: Penguin Books, 2003.

Lüpold, Stefan, et al. "How Sexual Selection Can Drive the Evolution of Costly Sperm Ornamentation." *Nature* 533, no. 7604 (May 26, 2016): 535–38. https://doi.org/10.1038/nature18005.

Luzar, N., and G. Gottsberger. "Flower Heliotropism and Floral Heating of Five Alpine Plant Species and the Effect on Flower Visiting in *Ranunculus montanus* in the Austrian Alps. *Arctic, Antarctic, and Alpine Research* 33 (2001): 93– 99.

Lynch, Zachary R., Todd A. Schlenke, and Jacobus C. de Roode. "Evolution of Behavioural and Cellular Defences against Parasitoid Wasps in the *Drosophila melanogaster* Subgroup." *Journal of Evolutionary Biology* 29, no. 5 (May 2016): 1016–29. https://doi.org/10.1111/jeb.12842.

Maák, István, et al. "Tool Selection during Foraging in Two Species of Funnel Ants." *Animal Behaviour* 123 (January 2017): 207–16. https://doi.org/10.1016/j.anbehav.2016.11.005.

MacNeal, David. *Bugged: The Insects Who Rule the World and the People Obsessed with Them*. New York: St. Martin's Press, 2017.

Magni, Paula A., et al. "Forensic Entomologists: An Evaluation of Their Status." *Journal of Insect Science* 13, no. 1 (January 1, 2013): 78. https://doi.org/10.1673/031.013.7801.

"Malaria Vaccine Loses Effectiveness over Several Years." *The Guardian*, June 30, 2016. https://guardian.ng/features/health/malaria-vaccine-loses-effectiveness-over-several-years/ (accessed August 25, 2020).

Manev, Hari, and Nikola Dimitrijevic. "Fruit Flies for Anti-Pain Drug Discovery." *Life Sciences* 76, no. 21 (April 8, 2005): 2403–7. https://doi.org/10.1016/j.lfs.2004.12.007.

Marent, Thomas, with Ben Morgan. *Rainforest*. New York: DK Publishing, 2006.

Margonelli, Lisa. *Underbug: An Obsessive Tale of Termites and Technology*. New York: Scientific American/Farrar, Straus and Giroux, 2018.

Marshall, Stephen A. *Flies: The Natural History and Diversity of Diptera*. Buffalo, NY: Firefly Books, 2012.

———. *Insects: Their Natural History and Diversity*. Buffalo, NY: Firefly Books, 2006.

Martín-Vega, Daniel, Aida Gómez-Gómez, and Arturo Baz. "The 'Coffin Fly' *Conicera tibialis* (Diptera: Phoridae) Breeding on Buried Human Remains after a Postmortem Interval of 18 Years." *Journal of Forensic Science* 56, no. 6 (July 2011): 1654–56. https://doi.org/10.1111/j.1556-4029.2011.01839.x.

Mason, Andrew C., Michael L. Oshinsky, and Ron R. Hoy. "Hyperacute Directional Hearing in a Microscale Auditory System." *Nature* 410, no. 6829 (April 5, 2001): 686–90. https://doi.org/10.1038/35070564.

Masterson, A. "Insects Smarter Than We Thought, Macquarie University Academics Say." *Sydney Morning Herald*, April 21, 2016.

McAlister, Erica. *The Secret Life of Flies*. Richmond Hill, Ontario: Firefly Books, 2017.

McClung, Chuck. "Study: Flies on Food Should Make You Drop Your Fork." *USA Today*, August 14, 2014, www.usatoday.com/story/news/nation/2014

/08/14/flies-health-hazard-orkin-study/14044947/ (accessed January 31, 2017).

McDonald, Dave J., and Johannes Jacobus Adriaan Van der Walt. "Observations on the Pollination of *Pelargonium tricolor*, Section Campylia (Geraniaceae)." *South African Journal of Botany* 58, no. 5 (October 1992): 386–92.

McGavin, George C. *Insects, Spiders and Other Terrestrial Arthropods*. London: Dorling Kindersley, 2000.

McKeever, Sturgis. "Observations of Corethrella Feeding on Tree Frogs (*Hyla*)." *Mosquito News* 37 (1977): 522–23.

———, and Frank E. French. "*Corethrella* (Diptera: Corethrellidae) of Eastern North America: Laboratory Life History and Field Responses to Anuran Calls." *Annals of the Entomological Society of America* 84, no. 5 (September 1991): 493–97. https://doi.org/10.1093/aesa/84.5.493.

McKeever, S., and W. Keith Hartberg. "An Effective Method for Trapping Adult Female *Corethrella* (Diptera: Chaoboridae)." *Mosquito News* 40, no. 1 (January 1980): 111–12.

McKenna, Duane D., et al. "Roadkill Lepidoptera: Implications of Roadways, Roadsides, and Traffic Rates for the Mortality of Butterflies in Central Illinois." *Journal of the Lepidopterists' Society* 55 (2001): 63–68.

McKie, Robin. "Europe at Risk from Spread of Tropical Insect-Borne Diseases." *The Guardian*, April 14, 2019. www.theguardian.com/science/2019/apr/14/tropical-insect-diseases-europe-at-risk-dengue-fever.

McLean, Jesse. "A Moth-er's Love." *Toronto Star*, August 24, 2019, A1, A8.

McNeill, Lizzy. "How Many Animals Are Born in the World Every Day?" *More or Less*, BBC Radio 4, June 11, 2018. www.bbc.com/news/science-environment-44412495.

Mead, Rebecca. "The Prophet of Dystopia." *The New Yorker*, April 17, 2017, 38–47.

Meiswinkel, Rudy, et al. *Infectious Diseases of Livestock*, Vol. 1: *Vectors: Culicoides spp*, 93–136. Cape Town: Oxford University Press, 2004.

Mery, Frédéric, et al. "Public Versus Personal Information for Mate Copying in an Invertebrate." *Current Biology* 19, no. 9 (May 12, 2009): 730–34. https://doi.org/10.1016/j.cub.2009.02.064.

Meuche, Ivonne, et al. "Silent Listeners: Can Preferences of Eavesdropping Midges Predict Their Hosts' Parasitism Risk?" *Behavioral Ecology* 27, no. 4 (July–August 2016): 995–1003. https://doi.org/10.1093/beheco/arw002.

Meve, U., and S. Liede. "Floral Biology and Pollination in Stapeliads—New Results and a Literature Review." *Plant Systematics and Evolution* 192 (1994): 99–116.

Meyer, Dagmar B., et al. "Development and Field Evaluation of a System to Collect Mosquito Excreta for the Detection of Arboviruses." *Journal of Medical Entomology* 56, no. 4 (July 2019): 1116–21. https://doi.org/10.1093/jme/tjz031.

Milan, Neil F., Balint Z. Kacsoh, and Todd A. Schlenke. "Alcohol Consumption as Self-Medication against Blood-Borne Parasites in the Fruit Fly." *Current Biology* 22, no. 6 (March 20, 2012): 488–93. https://doi.org/10.1016/j.cub.2012.01.045.

Miles, Ronald N., Daniel Robert, and Ron R. Hoy. "Mechanically Coupled Ears for Directional Hearing in the Parasitoid Fly *Ormia ochracea*." *The Journal of the Acoustical Society of America* 98, no. 6 (December 1995): 3059–70. https://doi.org/10.1121/1.413830.

Milius, Susan. "Long Tongue, Meet Short Flower." *ScienceNews*, September 5, 2015. https://www.sciencenews.org/article/long-tongued-fly-sips-afar (accessed January 31, 2017).

———. "Testing Mosquito Pee Could Help Track the Spread of Diseases." *ScienceNews*, April 5, 2019. https://www.sciencenews.org/article/testing-mosquito-pee-could-help-track-spread-diseases (accessed May 15, 2020).

Miller, Paige B., et al. "The Song of the Old Mother: Reproductive Senescence in Female *Drosophila*." *Fly* 8, no. 3 (December 18, 2014): 127–39. https://doi.org/10.4161/19336934.2014.969144.

Milton, Katherine. "Effects of Bot Fly (*Alouattamyia baeri*) Parasitism on a Free-Ranging Howler Monkey (*Alouatta palliata*) Population in Panama." *Journal of Zoology* 239, no. 1 (May 1996): 39–63. https://doi.org/10.1111/j.1469-7998.1996.tb05435.x.

Min, John, et al. "Harnessing Gene Drive." *Journal of Responsible Innovation* 5, supplement 1 (2018): S40–S65. https://doi.org/10.1080/23299460.2017.1415586.

Missagia, Caio C. C., and Maria Alice S. Alves. "Florivory and Floral Larceny by Fly Larvae Decrease Nectar Availability and Hummingbird Foraging Visits at Heliconia (Heliconiaceae) Flowers." *Biotropica* 49, no. 1 (January 2017): 13–17. https://doi.org/10.1111/btp.12368.

Möglich, Michael H. J., and Gary D. Alpert. "Stone Dropping by *Conomyrma bicolor* (Hymenoptera: Formicidae): A New Technique of Interference Competition." *Behavioral Ecology and Sociobiology* 6 (1979): 105–13. https://doi.org/10.1007/BF00292556.

Moré, Marcela, et al. "The Role of Fetid Olfactory Signals in the Shift to Saprophilous Fly Pollination in *Jaborosa* (Solanaceae)." *Arthropod-Plant Interactions* 13 (October 2018): 375–86. https://doi.org/10.1007/s11829-018-9640-y.

Moretti, Thiago de Carvalho, et al. "Insects on Decomposing Carcasses of Small Rodents in a Secondary Forest in Southeastern Brazil." *European Journal of Entomology* 105, no. 4 (October 2008): 691–96. www.eje.cz/scripts/viewabstract.php?abstract=1386.

Mortimer, Nathan. "Parasitoid Wasp Virulence: A Window into Fly Immunity." *Fly* 7, no. 4 (October 1, 2013): 242–48. https://doi.org/10.4161/fly.26484.

"Mosquitoes." National Geographic, n.d. https://www.nationalgeographic.com/animals/invertebrates/group/mosquitoes/ (accessed August 22, 2018).

Muth, Felicity. "Inside the Wonderful World of Bee Cognition—Where We're at Now." *Scientific American*, April 20, 2015. https://blogs.scientificamerican.com/not-bad-science/inside-the-wonderful-world-of-bee-cognition-where-we-re-at-now/.

Muto, L., et al. "An Innovative Ovipositor for Niche Exploitation Impacts Genital Coevolution between Sexes in a Fruit-Damaging *Drosophila*." *Proceedings of the Royal Society B* 285 (2018), 20181635.

Myers, Paul Z. "The Lovely Stalk-Eyed Fly." *ScienceBlogs*, March 15, 2007. http://scienceblogs.com/pharyngula/2007/03/15/the-lovely-stalkeyed-fly.

Nace, Trevor. "Morgan Freeman Converted His 124-Acre Ranch into a Giant Honeybee Sanctuary to Save the Bees." *Forbes*, March 20, 2019. www.forbes.com/sites/trevornace/2019/03/20/morgan-freeman-converted-his-124-acre-ranch-into-a-giant-honeybee-sanctuary-to-save-the-bees/#68b41857dfa5.

Naskrecki, Piotr. *The Smaller Majority*. Cambridge, MA: Belknap Press, 2005.

Nazni, Wasi Ahmad, et al. "First Report of Maggots of Family Piophilidae Recovered from Human Cadavers in Malaysia." *Tropical Biomedicine* 25, no. 2 (August 2008): 173–75.

Neckameyer, Wendi S., et al. "Dopamine and Senescence in *Drosophila melanogaster.*" *Neurobiology of Aging* 21, no. 1 (January–February 2000): 145–52.

New, Joshua J., and Tamsin C. German. "Spiders at the Cocktail Party: An Ancestral Threat That Surmounts Inattentional Blindness." *Evolution and Human Behavior* 36, no. 3 (August 2015): 165–73. https://doi.org/10.1016/j.evolhumbehav.2014.08.004.

The New York Times editorial board, "Insect Armageddon." *The New York Times*, October 29, 2017. www.nytimes.com/2017/10/29/opinion/insect-armageddon-ecosystem-.html (accessed May 15, 2020).

Newman, Barry. "Apple Turnover: Dutch Are Invading JFK Arrivals Building and None Too Soon—U.S.'s Best Known Airport Has Been a Lousy Place to Land, Walk, or Stand—Using Flies to Help Fliers." *The Wall Street Journal*, May 13, 1997, A1.

Nuñez Rodríguez, José, and Jonathan Liria. "Seasonal Abundance in Necrophagous Diptera and Coleoptera from Northern Venezuela." *Tropical Biomedicine* 34, no. 2 (June 2017): 315–23. https://www.researchgate.net/publication/317559366_Seasonal_abundance_in_necrophagous_Diptera_and_Coleoptera_from_northern_Venezuela.

Ofstad, Tyler A., Charles S. Zuker, and Michael B. Reiser. "Visual Place Learning in *Drosophila melanogaster.*" *Nature* 474 (2011): 204–7.

Oldroyd, Harold. "Dipteran." *Encyclopædia Britannica*, October 30, 2018 (mention of petroleum flies). www.britannica.com/animal/dipteran (accessed May 14, 2020).

Olotu, Ally, et al. "Seven-Year Efficacy of RTS,S/AS01 Malaria Vaccine among Young African Children." *The New England Journal of Medicine* 374, no. 26 (June 30, 2016): 2519–29. https://doi.org/10.1056/NEJMoa1515257.

Orford, Katherine A., Ian P. Vaughan, and Jane Memmott. "The Forgotten Flies: The Importance of Non-Syrphid Diptera as Pollinators." *Proceedings of the Royal Society B: Biological Sciences* 282, no. 1805 (April 22, 2015). https://doi.org/10.1098/rspb.2014.2934.

Owald, David, Suewei Lin, and Scott Waddell. "Light, Heat, Action: Neural Control of Fruit Fly Behaviour." *Philosophical Transactions of the Royal Society, B: Biological Sciences* 370, no. 1677 (September 19, 2015). https://doi.org/10.1098/rstb.2014.0211.

Paine, Robert T. "A Note on Trophic Complexity and Community Stability." *The American Naturalist* 103, no. 929 (January–February 1969): 91–93. https://doi.org/10.1086/282586.

Pape, Thomas, Daniel Bickel, and Rudolf Meier, eds. *Diptera Diversity: Status, Challenges and Tools*. Leiden: Brill, 2009.

Patterson, J. T., and W. S. Stone. *Evolution in the Genus Drosophila*. New York: Macmillan, 1952.

Patterson, K. David. "Yellow Fever Epidemics and Mortality in the United States, 1693–1905." *Social Science & Medicine* 34, no. 8 (April 1992): 855–65. https://doi.org/10.1016/0277-9536(92)90255-O (accessed May 15, 2020).

Paulson, Gregory S., and Eric R. Eaton. *Insects Did It First*. Xlibris, 2018.

Pearce, Fred. "Inventing Africa." *New Scientist* 167, no. 2251 (August 12, 2000): 30.

Pearson, Gwen. "50 Shades of Wrong: Disturbing Insect Sex." *Wired*, February 9, 2015. www.wired.com/2015/02/50-shades-wrong-disturbing-insect-sex / (accessed May 10, 2019).

Pennisi, Elizabeth. "This Fly Survives a Deadly Lake by Encasing Itself in a Bubble: Here's How It Makes It." *Science*, November 20, 2017. https://doi .org/10.1126/science.aar5258.

Perera, Hirunika, and Tharaka Wijerathna. "Sterol Carrier Protein Inhibition–Based Control of Mosquito Vectors: Current Knowledge and Future Perspectives." *Canadian Journal of Infectious Diseases and Medical Microbiology* 2019, no. 4 (July 2019): 1–6. https://doi.org/10.1155/2019/7240356.

Perry, Clint J., and Andrew B. Barron. "Honey Bees Selectively Avoid Difficult Choices." *Proceedings of the National Academy of Sciences of the United States of America* 110, no. 47 (November 19, 2013): 19155–59. https://doi.org /10.1073/pnas.1314571110.

———. "Neural Mechanisms of Reward in Insects." *Annual Review of Entomology* 58, no. 1 (September 2012): 543–62. https://doi.org/10.1146/annurev -ento-120811-153631.

Pierce, John D., Jr. "A Review of Tool Use in Insects." *Florida Entomologist* 69, no. 1 (March 1986): 95–104. https://doi.org/10.2307/3494748.

Policha, T., et al. "Disentangling Visual and Olfactory Signals in Mushroom Mimicking Dracula Orchids Using Realistic Three-Dimensional Printed Flowers." *New Phytologist* 210 (2016):1058–71.

Pomerleau, Mark. "AFRL Working on Insect-Eye View for Urban Targeting." Defense Systems, March 12, 2015. https://defensesystems.com/articles /2015/03/12/afrl-artificial-compound-eye-targeting.aspx.

Porter, Sanford D. "Biology and Behavior of *Pseudacteon* Decapitating Flies (Diptera: Phoridae) That Parasitize *Solenopsis* Fire Ants (Hymenoptera: Formicidae)." *Florida Entomologist* 81, no. 3 (September 1998): 292–309. https://doi.org/10.2307/3495920.

Preston-Mafham, Kenneth G. "Courtship and Mating in *Empis* (*Xanthempis*) *trigramma* Meig., *E. tessellata* F. and *E.* (*Polyblepharis*) *opaca* F. (Diptera: Empididae) and the Possible Implications of 'Cheating' Behaviour." *Journal of Zoology* 247, no. 2 (February 1999): 239–46. https://doi.org/10.1111/j.1469 -7998.1999.tb00987.x.

———. "Post-Mounting Courtship and the Neutralizing of Male Competitors Through 'Homosexual' Mountings in the Fly *Hydromyza livens* F. (Diptera: Scatophagidae)." *Journal of Natural History* 40, no. 1–2 (April 2006): 101–5. https://doi.org/10.1080/00222930500533658.

Pridgeon, A. M., et al. *Genera Orchidacearum*. Vol. 4: *Epidendroideae (Part 1)*. Oxford: Oxford University Press, 2005.

Puniamoorthy, Naline, Marion Kotrba, and Rudolf Meier. "Unlocking the 'Black Box': Internal Female Genitalia in Sepsidae (Diptera) Evolve Fast and Are Species-Specific." *BMC* [BioMed Central] *Evolutionary Biology* 10, article 275 (2010). https://doi.org/10.1186/1471-2148-10-275.

Puniamoorthy, Nalini, et al. "From Kissing to Belly Stridulation: Comparative Analysis Reveals Surprising Diversity, Rapid Evolution, and Much Homoplasy in the Mating Behaviour of 27 Species of Sepsid Flies (Diptera: Sepsidae)." *Journal of Evolutionary Biology* 22, no. 11 (November 2009): 2146–56. https://doi.org/10.1111/j.1420-9101.2009.01826.x.

Putz, Gabriele, and Martin Heisenberg. "Memories in *Drosophila* Heat-Box Learning." *Learning & Memory* 9, no. 5 (September 2002): 349–59. https://doi.org/10.1101/lm.50402.

Radonjić, Sanja, Snježana Hrnčić, and Tatjana Perović. "Overview of Fruit Flies Important for Fruit Production on the Montenegro Seacoast." *Biotechnology, Agronomy, Society and Environment* 23, no. 1 (2019): 46–56. https://doi.org/10.25518/1780-4507.17776.

Renner, Susanne S. "Rewardless Flowers in the Angiosperms, and the Role of Insect Cognition in Their Evolution." In *Plant-Pollinator Interactions: From Specialization to Generalization*, ed. Nickolas M. Waser and Jeff Ollerton, 123–44. Chicago: University of Chicago Press, 2006.

———, and Robert E. Ricklefs. "Dioecy and Its Correlates in the Flowering Plants." *American Journal of Botany* 82, no. 5 (May 1995): 596–606. https://doi.org/10.1002/j.1537-2197.1995.tb11504.x.

Rifai, Ryan. "UN: 200,000 Die Each Year from Pesticide Poisoning." *Al Jazeera*, March 8, 2017. www.aljazeera.com/news/2017/03/200000-die-year-pesticide-poisoning-170308140641105.html.

Rivers, David B., and Gregory A. Dahlem. *The Science of Forensic Entomology*. Chichester, UK: John Wiley & Sons, 2014.

Roberts, Andrew, et al. "Results from the Workshop 'Problem Formulation for the Use of Gene Drive in Mosquitoes.'" *The American Journal of Tropical Medicine and Hygiene* 96, no. 3 (March 8, 2017): 530–33. https://doi.org/10.4269/ajtmh.16-0726.

Robinson, Samuel V. J. "Plant-Pollinator Interactions at Alexandra Fiord, Nunavut." *Trail Six: An Undergraduate Geography Journal* 5 (2011): 13–20.

Rosenberg, Kenneth V., et al. "Decline of the North American Avifauna." *Science* 366, no. 6461 (October 4, 2019): 120–24. https://doi.org/10.1126/science.aaw1313.

Rowley, Wayne A., and Marcia Cornford. "Scanning Electron Microscopy of the Pit of the Maxillary Palp of Selected Species of *Culicoides*." *Canadian*

Journal of Zoology 50, no. 9 (September 1972): 1207–10. https://doi.org/10.1139/z72-162.

RTS,S Clinical Trials Partnership. "Efficacy and Safety of RTS,S/AS01 Malaria Vaccine with or without a Booster Dose in Infants and Children in Africa: Final Results of a Phase 3, Individually Randomised, Controlled Trial." *The Lancet* 386, no. 9988 (July 4, 2015): 31–45. https://doi.org/10.1016/S0140-6736(15)60721-8.

Runyon, Justin B., and Richard L. Hurley. "A New Genus of Long-Legged Flies Displaying Remarkable Wing Directional Asymmetry." *Proceedings of the Royal Society B: Biological Sciences* 271, supplement 3 (February 7, 2004): S114–16. https://doi.org/10.1098/rsbl.2003.0118.

Sánchez-Bayo, Francisco, and Kris A. G. Wyckhuys. "Worldwide Decline of the Entomofauna: A Review of Its Drivers." *Biological Conservation* 232 (April 2019): 8–27. https://doi.org/10.1016/j.biocon.2019.01.020.

Sansoucy, R. "Livestock—A Driving Force for Food Security and Sustainable Development." *World Animal Review* (FAO), 1995. http://www.fao.org/3/v8180t/v8180T07.htm (accessed August 21, 2020).

Sarkar, Sahotra. "Researchers Hit Roadblocks with Gene Drives." *BioScience* 68, no. 7 (July 2018): 474–80. https://doi.org/10.1093/biosci/biy060.

"Scientists Discover Why Flies Are So Hard to Swat." Phys.org, August 28, 2008. https://phys.org/news/2008-08-scientists-flies-hard-swat.html#jCp.

Scott, Jeffrey G., et al. "Insecticide Resistance in House Flies from the United States: Resistance Levels and Frequency of Pyrethroid Resistance Alleles." *Pesticide Biochemistry and Physiology* 107, no. 3 (November 2013): 377–84. https://doi.org/10.1016/j.pestbp.2013.10.006.

Scott, Kristin. Faculty Research Page, Department of Molecular and Cell Biology, University of California, Berkeley. Last updated January 1, 2019. https://mcb.berkeley.edu/faculty/NEU/scottk.html (accessed April 2019).

Scudder, Geoffrey G. E. "Comparative Morphology of Insect Genitalia." *Annual Review of Entomology* 16 (1971): 379–406. https://doi.org/10.1146/annurev.en.16.010171.002115.

Session, Laura A., and Steven D. Johnson. "The Flower and the Fly: Long Insect Mouthparts and Deep Floral Tubes Have Become So Specialized That Each Organism Has Become Dependent on the Other." *Natural History*, March 2005. www.naturalhistorymag.com/htmlsite/master.html?https://www.naturalhistorymag.com/htmlsite/0305/0305_feature.html.

Ševčík, Jan, Jostein Kjærandsen, and Stephen A. Marshall. "Revision of *Speolepta* (Diptera: Mycetophilidae), with Descriptions of New Nearctic and Oriental Species." *The Canadian Entomologist* 144, no. 1 (February 23, 2012): 93–107. https://doi.org/10.4039/tce.2012.10.

Shanor, Karen, and Jagmeet Kanwal. *Bats Sing, Mice Giggle: The Surprising Science of Animals' Inner Lives*. London: Icon Books, 2009.

Shao, Lisha, et al. "A Neural Circuit Encoding the Experience of Copulation in Female *Drosophila*." *Neuron* 102, no. 5 (June 5, 2019): 1025–36. https://doi.org/10.1016/j.neuron.2019.04.009.

Sheehan, Michael J., and Elizabeth A. Tibbetts. "Specialized Face Learning Is Associated with Individual Recognition in Paper Wasps." *Science* 334, no. 6060 (December 2, 2011): 1272–75. https://doi.org/10.1126/science.1211334.

Sheppard, D. Craig, et al. "A Value-Added Manure Management System Using the Black Soldier Fly." *Bioresource Technology* 50, no. 3 (1994): 275–79. https://doi.org/10.1016/0960-8524(94)90102-3.

Sherman, Ronald A., et al. "Maggot Therapy." In *Biotherapy: History, Principles and Practice: A Practical Guide to the Diagnosis and Treatment of Disease Using Living Organisms*, ed. M. Grassberger et al. Dordrecht, Netherlands: Springer Science+Business Media, 2013.

Shohat-Ophir, Galit, et al. "Sexual Deprivation Increases Ethanol Intake in *Drosophila*." *Science* 335, no. 6074 (March 16, 2012): 1351–55. https://doi.org/10.1126/science.1215932.

Shubin, Neal H., and Farish A. Jenkins, Jr. "An Early Jurassic Jumping Frog." *Nature* 377 (September 7, 1995), 49–52. https://doi.org/10.1038/377049a0.

Shuttlesworth, Dorothy. *The Story of Flies*. New York: Doubleday, 1970.

Sjöberg, Fredrik. *The Fly Trap*. New York: Vintage, 2015.

Smith, Ronald L. *Interior and Northern Alaska: A Natural History*. Bothell, WA: Book Publishers Network, 2008.

Sokolowski, Marla B. "Social Interactions in 'Simple' Model Systems." *Neuron* 65, no. 6 (March 25, 2010): 780–94. https://doi.org/10.1016/j.neuron.2010.03.007.

Spielman, Andrew, and Michael D'Antonio. *Mosquito: A Natural History of Our Most Persistent and Deadly Foe*. New York: Hyperion, 2002.

Ssymank, Axel, and Carol Kearns. "Flies–Pollinators on Two Wings." The New Diptera Site. http://diptera.myspecies.info/diptera/content/flies-pollinators-two-wings, 2009 (accessed August 7, 2020).

Ssymank, Axel, et al. *Das europäische Schutzgebietssystem NATURA 2000*. Schriftenreihe für Landschaftspflege und Naturschutz, vol. 53. Bonn–Bad Godesberg: Bundesamt für Naturschutz, 1998.

Stensmyr, Marcus C., et al. "Pollination: Rotting Smell of Dead-Horse Arum Florets." *Nature* 420 (2002): 625–26. https://doi.org/10.1038/420625a.

Stiling, Peter D. *Florida's Butterflies and Other Insects*. Sarasota, FL: Pineapple Press, 1989.

Stinson, Liz. "Enchanting Paintings Made from the Puke of 250,000 Flies." *Wired*, August 5, 2013. www.wired.com/2013/08/beautiful-abstract-paintings-made-from-the-puke-of-250000-flies/.

Sultan, Mehmet. "Forensic Entomology: How Insects Solve Murder Cases." *The Fountain* 53 (January–March 2006). https://fountainmagazine.com/2006/issue-53-january-march-2006/forensic-entomology-how-insects-solve-murder-cases (accessed August 18, 2020).

Sun, Dan, et al. "Progress and Prospects of CRISPR/Cas Systems in Insects and Other Arthropods." *Frontiers in Physiology* 8 (September 6, 2017): 608. https://doi.org/10.3389/fphys.2017.00608.

Sverdrup-Thygeson, Anne. *Buzz Sting Bite: Why We Need Insects*. New York: Simon & Schuster, 2019.

Syracuse University. "Forget Peacock Tails, Fruit Fly Sperm Tails Are the Most Extreme Ornaments: Syracuse University Researchers Among Those to Author New Paper in *Nature* That Explains Why Ornament May Have Evolved." *EurekAlert!*, May 25, 2016. www.eurekalert.org/pub_releases/2016-05/su-fpt052516.php.

Tabone, C. J., and J. S. de Belle. "Second-Order Conditioning in *Drosophila*." *Learning & Memory* 18, no. 4 (2011): 250–53.

Tarsitano, Michael S., and Robert R. Jackson. "Araneophagic Jumping Spiders Discriminate between Detour Routes That Do and Do Not Lead to Prey." *Animal Behaviour* 53, no. 2 (February 1997): 257–66. https://doi.org/10.1006/anbe.1996.0372.

Taylor, Barbara, Jen Green, and John Farndon. *The Big Bug Book*. London: Anness, 2004.

Taylor, David B., Roger D. Moon, and Darrell R. Mark. "Economic Impact of Stable Flies (Diptera: Muscidae) on Dairy and Beef Cattle Production." *Journal of Medical Entomology* 49, no. 1 (January 2012): 198–209. https://doi.org/10.1603/ME10050.

Teale, Edwin Way. *The Strange Lives of Familiar Insects*. New York: Dodd, Mead, 1964.

Theodor, Oskar. *On the Structure of the Spermathecae and Aedeagus in the Asilidae and Their Importance in the Systematics of the Family*. Jerusalem: Israel Academy of Sciences and Humanities, 1976.

Thompson, Christopher R., et al. "Bacterial Interactions with Necrophagous Flies." *Annals of the Entomological Society of America* 106, no. 6 (November 1, 2013): 799–809. https://doi.org/10.1603/AN12057.

Thornhill, Randy, and John Alcock. *The Evolution of Insect Mating Systems*. Cambridge, MA: Harvard University Press, 1983.

Tiffin, Helen. "Do Insects Feel Pain?" *Animal Studies Journal* 5, no. 1 (2016): 80–96. https://ro.uow.edu.au/asj/vol5/iss1/6.

Tiusanen, Mikko, et al. "One Fly to Rule Them All—Muscid Flies Are the Key Pollinators in the Arctic." *Proceedings of the Royal Society B: Biological Sciences* 283, no. 1839 (September 28, 2016). https://doi.org/10.1098/rspb.2016.1271.

Torres Toro, Juliana, et al. "An Update of Diversity of Soldier Flies (Stratiomyidae) from Colombia and Notes on Distribution in Colombian Biogeographical Provinces." Abstract 281, 9th International Congress of Dipterology, Windhoek, Namibia, 2018.

University of Michigan Health System. "Fruit Flies with Better Sex Lives Live Longer." *ScienceDaily*, November 28, 2013. www.sciencedaily.com/releases/2013/11/131128141258.htm.

Vandertogt, Alysha. "Can Mosquitoes and Black Flies Transmit COVID-19?" Cottage Life, May 19, 2020. https://cottagelife.com/general/can-mosquitoes-and-black-flies-transmit-covid-19/ (accessed August 21, 2020).

VanLaerhoven, Sherah L., and Ryan W. Merritt. "50Years Later, Insect Evidence Overturns Canada's Most Notorious Case—*Regina v. Steven Truscott*." *Forensic Science International* 301 (August 2019): 326–30. https://doi.org/10.1016/j.forsciint.2019.04.032.

Van Niekerken, Bill. "The Medfly Invasion: How a Tiny Insect Upended Bay Area Life Decades Ago." *San Francisco Chronicle*, September 19, 2017, updated November 25, 2018. www.sfchronicle.com/chronicle_vault/article/The-medfly-invasion-How-a-tiny-insect-upended-12205233.php (accessed May 15, 2020).

Van Swinderen, Bruno, and R. Andretic. "Dopamine in *Drosophila*: Setting Arousal Thresholds in a Miniature Brain." *Proceedings of Biological Science* 278 (2011): 906–13.

Vargas-Terán, Moisés, H. C. Hofmann, and N. E. Tweddle. "Impact of Screwworm Eradication Programmes Using the Sterile Insect Technique." In *Sterile Insect Technique: Principles and Practice in Area-Wide Integrated Pest Management*, ed. Victor Arnold Dyck, Jorge Hendrichs, and Alan S. Robinson, 629–50. New York: Springer, 2005.

Vinauger, Clément, et al. "Modulation of Host Learning in *Aedes aegypti* Mosquitoes." *Current Biology* 28, no. 3 (February 5, 2018): 333–44. https://doi.org/10.1016/j.cub.2017.12.015.

Vogel, Stephen. "Flickering Bodies: Floral Attraction by Movement." *Beiträge zur Biologie der Pflanzen* 72 (January 2001): 89–154.

Wake, Marvalee H. "Amphibian Locomotion in Evolutionary Time." *Zoology* 100 (1997): 141–51.

Waldbauer, Gilbert. *What Good Are Bugs? Insects in the Web of Life*. Cambridge, MA: Harvard University Press, 2003.

Wandersee, James H., and Elisabeth E. Schussler. "Preventing Plant Blindness." *American Biology Teacher* 61 (1999): 82–86.

Wangberg, James K. *Six-Legged Sex: The Erotic Lives of Bugs*. Golden, CO: Fulcrum, 2001.

Wee, Suk Ling, Shwu Bing Tana, and Andreas Jürgens. "Pollinator Specialization in the Enigmatic *Rafflesia cantleyi*: A True Carrion Flower with Species-Specific and Sex-Biased Blow Fly Pollinators." *Phytochemistry* 153 (September 2018): 120–28. https://doi.org/10.1016/j.phytochem.2018.06.005.

Weeks, Emma N. I., et al. "Effects of Four Commercial Fungal Formulations on Mortality and Sporulation in House Flies (*Musca domestica*) and Stable Flies (*Stomoxys calcitrans*)." *Medical and Veterinary Entomology* 31, no. 1 (March 2017): 15–22. https://doi.org/10.1111/mve.12201.

Weisberger, Mindy. "How Much Do You Poop in Your Lifetime?" *LiveScience*, March 21, 2018. www.livescience.com/61966-how-much-you-poop-in-lifetime.html (accessed May 15, 2020).

Weiss, Harry B. "Insects and Pain." *The Canadian Entomologist* 46, no. 8 (August 1914): 269–71. https://doi.org/10.4039/Ent46269-8.

Welsh, Jennifer. "World's Tiniest Fly May Decapitate Ants, Live in Their Heads." *Live Science*, July 2, 2012. www.livescience.com/21326-smallest-fly-decapitates-ants.html.

Wheeler, Quentin. "New to Nature No 88: *Euryplatea nanaknihali*: A Parasitoid Discovered in Thailand Is the World's Smallest Fly." *The Guardian*, October 13, 2012. www.theguardian.com/science/2012/oct/14/euryplatea-nanaknihali-new-to-nature.

Whitman, William B., David C. Coleman, and William J. Wiebe. "Prokaryotes: The Unseen Majority." *Proceedings of the National Academy of Sciences of the United States of America* 95, no. 12 (June 9, 1998): 6578–83. https://doi.org/10.1073/pnas.95.12.6578.

Whitworth, Terry L. Blow Flies home page. http://www.blowflies.net/ (accessed July 22, 2019).

Wigglesworth, Vincent B. "Do Insects Feel Pain?" *Antenna* 4 (1980): 8–9.

Winegard, Timothy C. *The Mosquito: A Human History of Our Deadliest Predator*. New York: Dutton, 2019.

Witze, Alexandra. "Flying Insects Tell Tales of Long-Distance Migrations: Well-Timed Travel Ensures Food and Breeding Opportunities." *Science News*, April 5, 2018. www.sciencenews.org/article/flying-insects-tell-tales-long-distance-migrations?utm_source=email&utm_medium=email&utm_campaign=latest-newsletter-v2.

Wohlleben, Peter. *The Inner Life of Animals: Love, Grief, and Compassion—Surprising Observations of a Hidden World*. Vancouver: Greystone Books, 2017.

World Health Organization. "Malaria: Insecticide Resistance." Last updated February 19, 2020. www.who.int/malaria/areas/vector_control/insecticide_resistance/en (accessed May 15, 2020).

———. "The Top 10 Causes of Death." May 24, 2018. www.who.int/news-room/fact-sheets/detail/the-top-10-causes-of-death.

———. *Vector Resistance to Pesticides: Fifteenth Report of the WHO Expert Committee on Vector Biology and Control* [meeting held in Geneva from 5 to 12 March 1991]. Geneva: World Health Organization, 1992. https://apps.who.int/iris/handle/10665/37432.

Wu, Xinwei, and Shucun Sun. "The Roles of Beetles and Flies in Yak Dung Removal in an Alpine Meadow of Eastern Qinghai-Tibetan Plateau." *Écoscience* 17, no. 2 (June 2010): 146–55. https://doi.org/10.2980/17-2-3319.

Wulf, Andrea. *The Invention of Nature: Alexander von Humboldt's New World*. New York: Vintage, 2016.

Yarali, Ayse, et al. "'Pain Relief' Learning in Fruit Flies." *Animal Behaviour* 76, no. 4 (October 2008): 1173–85. https://doi.org/10.1016/j.anbehav.2008.05.025.

Yin, Jerry C. P., et al. "CREB as a Memory Modulator: Induced Expression of a dCREB2 Activator Isoform Enhances Long-Term Memory in *Drosophila*." *Cell* 81, no. 1 (April 7, 1995): 107–15. https://doi.org/10.1016/0092-8674(95)90375-5.

Yong, Ed. "Scientists Genetically Engineered Flies to Ejaculate Under Red Light." *The Atlantic*, April 19, 2018. www.theatlantic.com/science/archive/2018/04/scientists-genetically-engineered-flies-to-ejaculate-under-red-light/558320/.

Young, Allen M. *The Chocolate Tree: A Natural History of Cacao*, rev. ed. Gainesville: University Press of Florida, 2007.

Yurkovic, Alexandra, et al. "Learning and Memory Associated with Aggression in *Drosophila melanogaster*." *Proceedings of the National Academy of Sciences of the United States of America* 103, no. 46 (November 14, 2006): 17519–24. https://doi.org/10.1073/pnas.0608211103.

Zahavi, Amotz. "Mate Selection—a Selection for a Handicap." *Journal of Theoretical Biology* 53, no. 1 (September 1975): 205–14. https://doi.org/10.1016/0022-5193(75)90111-3.

Zeldovich, Lina. "New Study Finds Insects Speak in Different 'Dialects.'" *JSTOR Daily*, July 31, 2018. https://daily.jstor.org/new-study-finds-insects-speak-in-different-dialects (accessed May 15, 2020).

Zimmer, Carl. *Parasite Rex: Inside the Bizarre World of Nature's Most Dangerous Creatures*. New York: Free Press, 2000.

———. "These Animal Migrations Are Huge—and Invisible." *The New York Times*, June 13, 2019. https://www.nytimes.com/2019/06/13/science/animals-migration-insects.html.

Zivkovic, Bora. "Stumped by Bed Nets, Mosquitoes Turn Midnight Snack into Breakfast." *Scientific American*, October 3, 2012. https://blogs.scientificamerican.com/a-blog-around-the-clock/stumped-by-bed-nets-mosquitoes-turn-midnight-snack-into-breakfast/.

Zlomislic, Diana. "Fields of Dreams." *The Star* (Toronto), June 20, 2019. https://projects.thestar.com/climate-change-canada/saskatchewan/ (accessed August 4, 2020).

Zwarts, Liesbeth, Marijke Versteven, and Patrick Callaerts. "Genetics and Neurobiology of Aggression in *Drosophila*." *Fly* 6, no. 1 (January–March 2012): 35–48. https://doi.org/10.4161/fly.19249.

Index

Page numbers in italics refer to illustrations.